P9-CRB-852

STRUCTURAL VIBRATION ANALYSIS:
Modelling, Analysis and Damping of Vibrating Structures

ELLIS HORWOOD SERIES IN ENGINEERING SCIENCE

STRENGTH OF MATERIALS
J. M. ALEXANDER, University College of Swansea.
TECHNOLOGY OF ENGINEERING MANUFACTURE
J. M. ALEXANDER, R. C. BREWER, Imperial College of Science and Technology, University of London, J. R. CROOKALL, Cranfield Institute of Technology.
COMPUTER AIDED DESIGN AND MANUFACTURE 2nd Edition
C. B. BESANT, Imperial College of Science and Technology, University of London.
STRUCTURAL DESIGN AND SAFETY
D. I. BLOCKLEY, University of Bristol.
BASIC LUBRICATION THEORY 3rd Edition
ALASTAIR CAMERON, Imperial College of Science and Technology, University of London.
STRUCTURAL MODELLING AND OPTIMIZATION
D. G. CARMICHAEL, University of Western Australia
ADVANCED MECHANICS OF MATERIALS 2nd Edition
Sir HUGH FORD, F.R.S., Imperial College of Science and Technology, University of London and J. M. ALEXANDER, University College of Swansea.
ELASTICITY AND PLASTICITY IN ENGINEERING
Sir HUGH FORD, F.R.S. and R. T. FENNER, Imperial College of Science and Technology, University of London.
INTRODUCTION TO LOADBEARING BRICKWORK
A. W. HENDRY, B. A. SINHA and S. R. DAVIES, University of Edinburgh
ANALYSIS AND DESIGN OF CONNECTIONS BETWEEN STRUCTURAL JOINTS
M. HOLMES and L. H. MARTIN, University of Aston in Birmingham
TECHNIQUES OF FINITE ELEMENTS
BRUCE M. IRONS, University of Calgary, and S. AHMAD, Bangladesh University of Engineering and Technology, Dacca.
FINITE ELEMENT PRIMER
BRUCE IRONS and N. SHRIVE, University of Calgary
PROBABILITY FOR ENGINEERING DECISIONS: A Bayesian Approach
I. J. JORDAAN, University of Calgary
STRUCTURAL DESIGN OF CABLE-SUSPENDED ROOFS
L. KOLLAR, City Planning Office, Budapest and K. SZABO, Budapest Technical University.
CONTROL OF FLUID POWER, 2nd Edition
D. McCLOY, The Northern Ireland Polytechnic and H. R. MARTIN, University of Waterloo, Ontario, Canada.
TUNNELS: Planning, Design, Construction — Vols. 1 and 2
T. M. MEGAW and JOHN BARTLETT, Mott, Hay and Anderson, International Consulting Engineers
UNSTEADY FLUID FLOW
R. PARKER, University College, Swansea
DYNAMICS OF MECHANICAL SYSTEMS 2nd Edition
J. M. PRENTIS, University of Cambridge.
ENERGY METHODS IN VIBRATION ANALYSIS
T. H. RICHARDS, University of Aston, Birmingham.
ENERGY METHODS IN STRESS ANALYSIS: With an Introduction to Finite Element Techniques
T. H. RICHARDS, University of Aston, Birmingham.
ROBOTICS AND TELECHIRICS
M. W. THRING, Queen Mary College, University of London
STRESS ANALYSIS OF POLYMERS 2nd Edition
J. G. WILLIAMS, Imperial College of Science and Technology, University of London.

STRUCTURAL VIBRATION ANALYSIS:
Modelling, Analysis and Damping of Vibrating Structures

C. F. BEARDS, BSc., Ph.D., C.Eng., M.R.Ae.S.
Department of Mechanical Engineering
Imperial College of Science and Technology
University of London

ELLIS HORWOOD LIMITED
Publishers · Chichester

Halsted Press: a division of
JOHN WILEY & SONS
New York · Brisbane · Chichester · Toronto

First published in 1983 by

ELLIS HORWOOD LIMITED
Market Cross House, Cooper Street, Chichester, West Sussex, PO19 1EB, England

The publisher's colophon is reproduced from James Gillison's drawing of the ancient Market Cross, Chichester.

Distributors:

Australia, New Zealand, South-east Asia:
Jacaranda-Wiley Ltd., Jacaranda Press,
JOHN WILEY & SONS INC.,
G.P.O. Box 859, Brisbane, Queensland 40001, Australia

Canada:
JOHN WILEY & SONS CANADA LIMITED
22 Worcester Road, Rexdale, Ontario, Canada.

Europe, Africa:
JOHN WILEY & SONS LIMITED
Baffins Lane, Chichester, West Sussex, England.

North and South America and the rest of the world:
Halsted Press: a division of
JOHN WILEY & SONS
605 Third Avenue, New York, N.Y. 10016, U.S.A.

British Library Cataloguing in Publication Data
Beards, C. F.
Structural vibration analysis. –
(Ellis Horwood series in engineering science)
1. Structural dynamics 2. Vibration
I. Title
624.1'71 TA654

Library of Congress Card No. 82-23299

ISBN 0-85312-325-X (Ellis Horwood Ltd., Publishers – Library Edn.)
ISBN 0-85312-579-1 (Ellis Horwood Ltd., Publishers – Student Edn.)
ISBN 0-470-27422-0 (Halsted Press)

Printed in Great Britain by Unwin Brothers, of Woking.

Contents

Preface

There are many books, codes of practice and other publications on the design of structures subjected to static loads, but there has been little published work to assist the designer with predicting the effects of dynamic loading on structures. There is information available on how to test structures under dynamic loads and what to measure, but these are of little help to the designer wanting to produce a structure with a particular dynamic performance.

Excellent advanced specialised texts on the vibration analysis of dynamic systems are available, and some are referred to later; but they require advanced mathematical knowledge and understanding of dynamic problems, and often refer to idealised systems rather than mathematical models of structures. This text links basic vibration analysis with these advanced texts; it is not only an introduction to advanced analysis methods but also discusses the theory relevant to structural vibration analysis. At the same time it describes how structural parameters can be changed to achieve a desired dynamic performance, and the mechanisms and control of structural damping.

This book is intended to give practising engineers and designers, and students of engineering to first degree level, a thorough understanding of the principles involved in the analysis of structural vibration, and to provide a sound theoretical basis for further study. It should also enable the designer to predict the effect of dynamic loads on structures, and to alter a structure so that a particular dynamic performance is achieved. A number of worked examples, and a selection of problems, have been included.

August 1982
C. F. Beards

General notation

a	damping factor, dimension.
b	circular frequency (rad/s), dimension.
c	coefficient of viscous damping, velocity of propagation of stress wave $= \sqrt{E/\rho}$.
c_c	coefficient of critical viscous damping $= 2\sqrt{mk}$.
c_d	equivalent viscous damping coefficient for dry friction damping $= 4F_d/\pi\omega X$.
c_H	equivalent viscous damping coefficient for hysteretic damping $= \eta k/\omega$.
f	frequency (Hz), exciting force.
f_s	Strouhal frequency (Hz).
h	height, thickness.
j	$\sqrt{-1}$.
k	linear spring stiffness, beam shear constant.
k_T	torsional spring stiffness.
k^*	complex stiffness $= k(1 + j\eta)$.
l	length.
m	mass.
q	generalised coordinate.
s	Laplace operator $= a + jb$.
t	time.
u	displacement.
v	velocity, deflection.
x	displacement.
y	displacement.
z	displacement.
A	amplitude, constant, cross-sectional area.

B	constant.
$C_{1,2,3,4}$	constants.
D	flexural rigidity $= Eh^3/12(1-\nu^2)$, hydraulic mean diameter.
E	modulus of elasticity.
E'	in-phase, or storage modulus.
E''	quadrature, or loss modulus.
E^*	complex modulus $= E' + jE''$.
F	exciting force amplitude.
F_d	coulomb friction force $= \mu N$.
F_T	transmitted force.
G	centre of mass, modulus of rigidity.
I	mass moment of inertia.
J	second moment of area.
K	stiffness.
M	mass, moment, mobility.
N	applied normal force.
P	force.
Q	factor of damping $= 1/2\zeta$.
R	radius of curvature.
T	kinetic energy, tension.
T_R	transmissibility $= F_T/F$.
V	potential energy.
X	amplitude of motion.
X_S	static deflection $= F/k$.
X/X_S	dynamic magnification factor.
Z	impedance.
α	coefficient, influence coefficient, phase angle, receptance.
β	coefficient, receptance.
γ	coefficient, receptance.
ϵ	short time, strain.
ϵ_0	strain amplitude.
η	loss factor $= E''/E'$.
ζ	damping ratio $= c/c_c$.
θ	angular displacement.
λ	matrix eigenvalue, $[\rho A\omega^2/EI]^{1/4}$.
μ	coefficient of friction, mass ratio $= m/M$.

ν	Poisson's ratio,
	circular exciting frequency (rad/s).
ρ	material density.
σ	stress.
σ_0	stress amplitude.
τ	period of vibration = 1/f.
τ_v	period of viscous damped vibration.
ϕ	phase angle,
	function of time.
ψ	phase angle.
ω	undamped circular frequency (rad/s).
ω_v	viscous damped circular frequency = $\omega\sqrt{1-\zeta^2}$.
Λ	logarithmic decrement = $\ln X_1/X_2$.
Ω	natural circular frequency (rad/s).

Chapter 1

Introduction

A structure is a supporting framework which may be part of a building, ship, space vehicle, engine or some other system; it can be a collection of elements fastened together or a single component. Before the Industrial Revolution started, structures usually had a very large mass because heavy timbers, castings and stonework were used in their fabrication; also the vibration excitation sources were small in magnitude so that the dynamic response of structures was extremely low. Furthermore these constructional methods usually produced a structure with very high inherent damping, which also gave a low structural response to dynamic excitation. Over the last two hundred years, with the advent of relatively strong lightweight materials such as cast iron, steel and aluminium, and increased knowledge of the material properties and structural loading, the mass of structures built to fulfil a particular function has decreased. The efficiency of engines has improved and, with higher rotational speeds, the magnitude of the vibration exciting forces has increased. This process of increasing excitation with reducing structural mass and damping has continued at an increasing pace to the present day when few, if any, structures can be designed without carrying out the necessary vibration analysis, if their dynamic performance is to be acceptable. It is therefore essential to carry out a vibration analysis of any proposed structure.

It is usually much easier to analyse and modify a structure at the design stage than it is to modify a structure with undesirable vibration characteristics after it has been built. However, it is sometimes necessary to be able to reduce the vibration of existing structures brought about by inadequate initial design, by change of function of the structure or by a change in the environmental conditions, and therefore techniques for the analysis of structural vibration should be applicable to existing structures as well as to those in the design stage. It is the solution to vibration problems which may be different depending on whether or not the structure exists.

To summarise, present-day structures often contain high energy sources which create intense vibration excitation problems, and modern construction methods result in structures with low mass and low inherent damping. Therefore careful design and analysis is necessary to avoid resonance or an undesirable dynamic performance.

1.1 THE CAUSES OF STRUCTURAL VIBRATION

There are two factors which control the amplitude and frequency of vibration in a structure; the excitation applied and the response of the structure to that particular excitation. Changing either the excitation or the dynamic characteristics of the structure will change the vibration stimulated.

The excitation arises from external sources such as ground or foundation vibration, cross winds, waves and currents, and earthquakes, and sources internal to the structure such as moving loads and rotating or reciprocating engines and machinery. These excitation forces and motions can be periodic or harmonic in time, or due to shock or impulse loadings, or even random in nature. All these types of excitation are considered in this text except random loading, because specialised mathematical techniques are required for random vibration analysis and these are adequately considered in existing texts, such as *An Introduction to Random Vibration and Spectral Analysis* by D. E. Newland (Longman).

The response of the structure to excitation depends on the method of applying, and the location of, the exciting force or motion, and the dynamic characteristics of the structure such as its natural frequencies and inherent damping level.

1.2 THE EFFECTS OF STRUCTURAL VIBRATION

In some structures, such as vibratory conveyors and compactors, vibration is encouraged, but these are special cases and in most structures vibration is undesirable. This is because vibration creates dynamic stresses and strains which can cause fatigue and failure of the structure, fretting corrosion between contacting elements, and noise in the environment; also it can impair the function and life of the structure or its components.

1.3 THE REDUCTION OF STRUCTURAL VIBRATION

The level of vibration in a structure can be attenuated by reducing either the excitation or the response of the structure to that excitation, or both. It is sometimes possible, at the design stage, to reduce the exciting force or motion by changing the equipment responsible, by relocating it within the structure, or by isolating it from the structure so that the generated vibration is not transmitted to the supports. The structural response can be altered by changing the mass or stiffness of the structure, by moving the source of excitation to another location, or by increasing the damping in the structure. Naturally, careful analysis is necessary to predict all the effects of any such changes, whether at the design stage or as a modification to an existing structure.

1.4 THE ANALYSIS OF STRUCTURAL VIBRATION

It is necessary to analyse the vibration of structures in order to predict the natural frequencies and the response to the expected excitation. The natural frequencies of the structure must be found because if the structure is excited at one of these frequencies resonance occurs, with resulting high vibration amplitudes, dynamic stresses and noise levels. Accordingly resonance should be avoided and the structure designed so that it is not encountered during normal conditions; this often means that the structure need only be analysed over the expected frequency range of excitation.

Although it may be possible to analyse the complete structure, this often leads to a very complicated analysis and the production of much unwanted information. A simplified mathematical model of the structure is usually sought therefore that will, when analysed, produce the desired information as economically as possible and with acceptable accuracy. The derivation of a simple mathematical model to represent the dynamics of a real structure is not easy, if the model is to produce useful and realistic information. It is often desirable for the model to predict the location of nodes in the structure. These are points of zero vibration amplitude and are thus useful locations for the siting of particularly delicate equipment. Also a particular mode of vibration cannot be excited by forces applied at one of its nodes.

1.5 THE MODELLING OF DYNAMIC STRUCTURES

All real structures possess an infinite number of degrees of freedom; that is, an infinite number of coordinates are necessary to specify completely the position of the structure at any instant of time. A structure possesses as many natural frequencies as it has degrees of freedom, and if excited at any of these natural frequencies a state of resonance exists, so that a large amplitude vibration response occurs. For each natural frequency the structure has a particular way of vibrating so that it has a characteristic shape, or mode of vibration, at each natural frequency.

Fortunately it is not usually necessary to calculate all the natural frequencies of a structure; this is because many of these frequencies will not be excited and in any case they may give small resonance amplitudes because the damping is high for that particular mode of vibration. Therefore the analytical model of a dynamic structure need have only a few degrees of freedom, or even only one, providing the structural parameters are chosen so that the correct mode of vibration is modelled. It is never easy to derive a realistic and useful mathematical model of a structure, because the analysis of particular modes of vibration is usually sought, and the determination of the relevant structural motions and parameters for the mathematical model needs great care. Some examples of models derived for real structures are given below, whilst further examples are given throughout the text.

The swaying oscillation of a chimney or tower can be investigated by means of a single degree of freedom model. This model would consider the chimney to be a rigid body resting on an elastic soil. To consider bending vibration in the chimney itself would require a more refined model such as the four degree of freedom system shown in Fig. 1.1. By giving suitable values to the mass and stiffness parameters a good approximation to the first bending mode frequency, and the corresponding mode shape, may be obtained. Such a model would not be sufficiently accurate for predicting the frequencies of higher modes; to accomplish this a more refined model with more mass elements and therefore more degrees of freedom would be necessary.

Vibrations of a machine tool can be analysed by modelling the machine

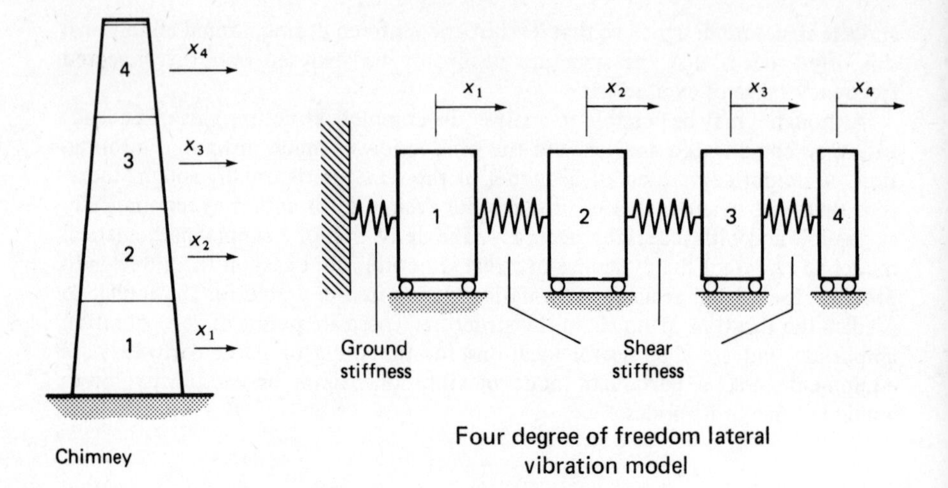

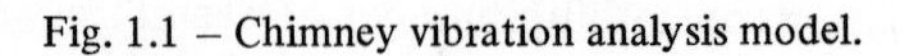

Fig. 1.1 – Chimney vibration analysis model.

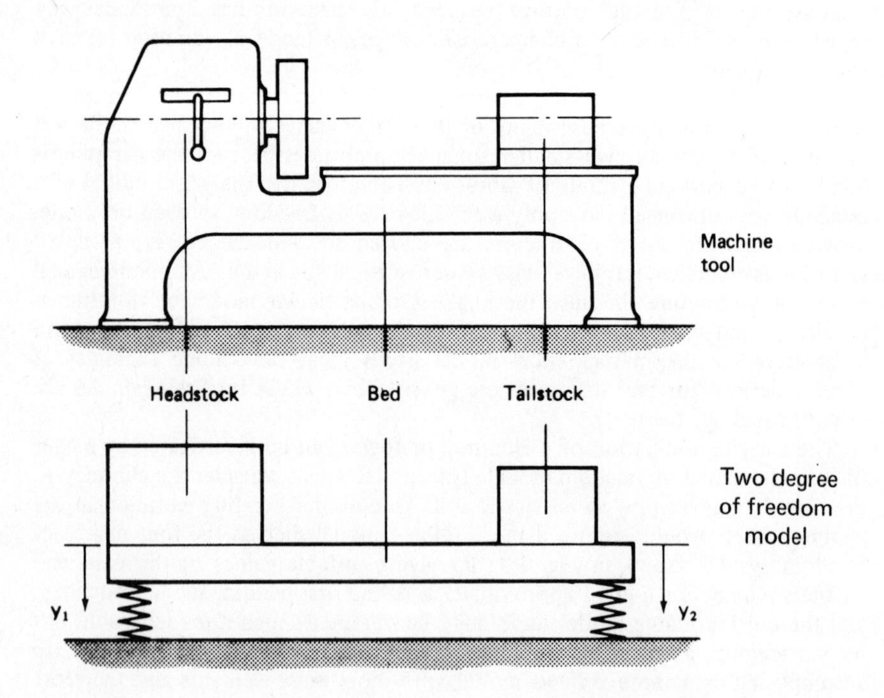

Fig. 1.2 – Machine tool vibration analysis model.

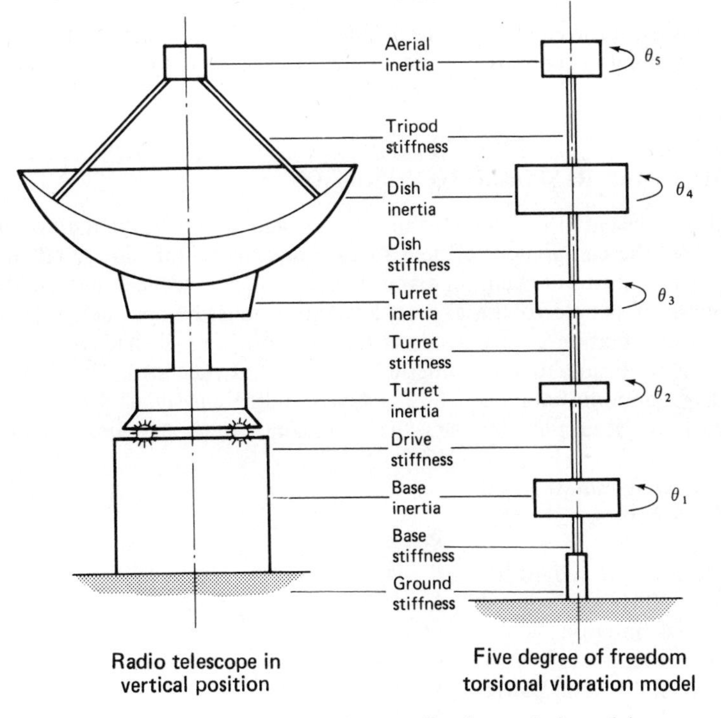

Fig. 1.3 – Radio telescope vibration analysis model.

structure by the two degree of freedom system shown in Fig. 1.2. In the simplest analysis the bed can be considered to be a rigid body with mass and inertia, and the headstock and tailstock are each modelled by lumped masses. The bed is supported by springs at each end as shown. Such a model would be useful for determining the lowest or fundamental natural frequency of vibration. A refinement to this model, which may be essential in some designs of machine where the bed cannot be considered rigid, is to consider the bed to be a flexible beam with lumped masses attached as before.

To analyse the torsional vibration of a radio telescope when in the vertical position a five degree of freedom model, as shown in Fig. 1.3, can be used. The mass and inertia of the various components may usually be estimated fairly accurately, but the calculation of the stiffness parameters at the design stage may be difficult; fortunately the natural frequencies are proportional to the square root of the stiffness. If the structure, or a similar one, is already built, the stiffness parameters can be measured. A further simplification of the model would be to put the turret inertia equal to zero, so that a three degree of freedom model is obtained. Such a model would be easy to analyse and would predict the lowest natural frequency of torsional vibration with fair accuracy, providing the correct inertia and stiffness parameters were used. It could not be used for predicting any other modes of vibration because of the coarseness of the model. However, in many structures only the lowest natural frequency is required, since if the structure can survive this frequency it will be able to survive other natural frequencies too.

None of these models include the effect of damping in the structure. Damping in most structures is very low so that the difference between the undamped and the damped natural frequencies is negligible. It is only necessary to include the effect of damping in the model if the response to a specific excitation is sought, particularly at frequencies in the region of a resonance.

1.6 THE HUMAN RESPONSE TO VIBRATION

Although it is essential for the dynamic stresses and strains in a structure to be withstood by the components of the structure, and that failure due to fatigue or malfunctioning must not occur, in many structures such as vehicles and buildings the response of people to the expected vibration must be considered. Human perception of vibration is very good, so that it is often a real challenge in structural design to ensure that the threshold level is not exceeded. The threshold for sensing harmonic vibration, both when standing and lying down, can be predicted fairly accurately by using the Diekmann criteria K-values as follows:

The Diekmann K-values

Vertical vibration:

- below 5 Hz $K = Af^2$
- between 5 Hz and 40 Hz $K = Af$
- above 40 Hz $K = 200A$

Horizontal vibration:

- below 2 Hz $K = 2Af^2$
- between 2 Hz and 25 Hz $K = 4Af$
- above 25 Hz $K = 100A$

where A = amplitude of vibration in mm, and f = frequency in Hz.

The regions for vibration sensitivity are as follows:

- $K = 0.1$, lower limit of perception.
- $K = 1$, allowable in industry for any period of time.
- $K = 10$, allowable for short duration only.
- $K = 100$, upper limit of strain for the average man.

Figure 1.4 shows approximate human vibration tolerance levels together with the Diekmann criteria for $K = 1$ and vertical vibration. It can be seen that $K = 0.1$ is a very safe value for predicting the perception threshold, which is the vibration level which should not be exceeded in buildings.

1.7 OUTLINE OF THE TEXT

The analysis of damped and undamped, free and forced vibration of single degree of freedom structures and models is described in Chapter 2. Not only does this analysis allow a wide range of problems to be solved, but it also is essential background to the multi-degree of freedom structure analysis considered in Chapter 3. Structures such as beams and plates with distributed mass and elasticity are analysed in Chapter 4. The damping which occurs in structures and its effect on structural response is described in Chapter 5, together with measurement and analysis techniques for damped structures, and methods for increasing the damping in structures. These chapters contain a number of worked examples to aid the understanding of the techniques described. Chapter 6 is devoted entirely to problems.

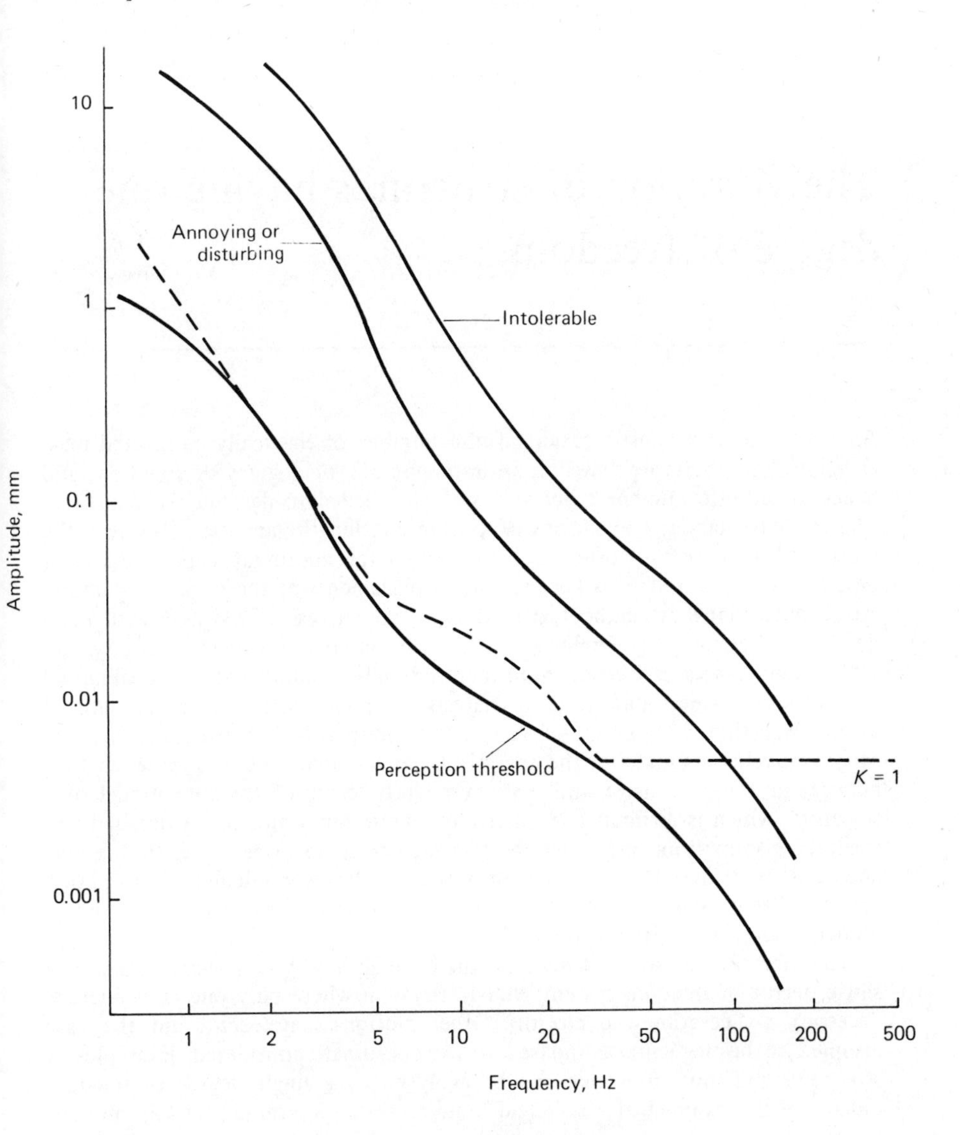

Fig. 1.4 – Human vibration tolerance.

Methods of modelling and analysis, including computer methods of solution are presented without becoming embroiled in computational detail. It must be stressed that the principles and analysis methods of any computer program used should be thoroughly understood before its application to any vibration problem. Round-off errors and other approximations may invalidate the results for the structure being analysed.

Chapter 2

The vibration of structures having one degree of freedom

All real structures consist of an infinite number of elastically connected mass elements and therefore have an infinite number of degrees of freedom; and hence an infinite number of coordinates are needed to describe their motion. This leads to elaborate equations of motion and lengthy analyses. However, the motion of a structure is often such that only a few coordinates are necessary to describe its motion. This is because the displacements of the other coordinates are so small that they can be neglected. Now, the analysis of a system with a few degrees of freedom is generally easier to carry out than the analysis of a system with many degrees of freedom, and therefore only a simple mathematical model of a structure is desirable from an analysis viewpoint. Although the amount of information that a simple model can yield is limited, if it is sufficient then the simple model is adequate for the analysis. Often a compromise has to be reached, between a comprehensive and elaborate multi-degree of freedom model of a structure, which is difficult and costly to analyse but yields much detailed and accurate information, and a simple few degrees of freedom model that is easy and cheap to analyse but yields less information. However, adequate information about the vibration of a structure can often be gained by analysing a simple model, at least in the first instance.

The vibration of some structures can be analysed by considering them as a **single degree of freedom system**; that is a system where only one coordinate is necessary to describe the motion. Other motions may occur, but they are assumed to be negligible compared to the coordinate considered. Examples of structures and motions which can be analysed by a single degree of freedom model are the swaying of a tall rigid building resting on an elastic soil, and the transverse vibration of a bridge. Before considering these examples in more detail, it is necessary to review the analysis of vibration of single degree of freedom dynamic systems. For a more comprehensive study see *Vibration Analysis and Control System Dynamics* by C. F. Beards (Ellis Horwood, 1981).

2.1 FREE UNDAMPED VIBRATION

2.1.1 Translation vibration

In the system shown in Fig. 2.1, a rigid body of mass m is free to move along a fixed horizontal surface. A spring of constant stiffness k which is fixed at one

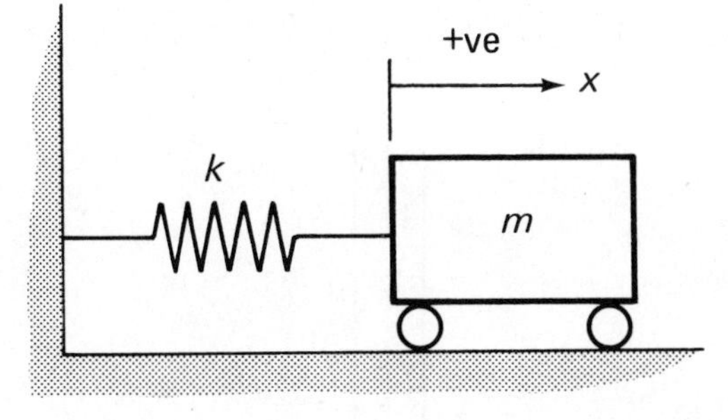

Fig. 2.1 – Single degree of freedom model – translation vibration.

end is attached at the other end to the body. The displacement of the body from the equilibrium position is given by x. If the body is displaced, it will vibrate in the x direction about its equilibrium position. The free-body diagrams (FBDs) for a displacement x in the positive direction shown are given in Figs. 2.2(a) and (b).

Fig. 2.2(a) – Applied force. (b) Effective force.

The equation of motion is, therefore

$$m\ddot{x} = -kx \qquad \text{or} \qquad \ddot{x} + \frac{k}{m}x = 0$$

and the motion of the body is simple harmonic, frequency $1/2\pi . \sqrt{k/m}$ Hz. A similar result is obtained from the analysis of a system where the body is supported to vibrate in the vertical direction only.

The general solution to the equation of motion is

$$x = A \cos \omega t + B \sin \omega t,$$

where A and B are constants which can be found by considering the initial conditions, and ω is the circular frequency of the motion, $\sqrt{k/m}$ rad/s.

2.1.2 Torsional vibration

Figure 2.3 shows the model used to study torsional vibration. A body with mass moment of inertia I about the axis of rotation is fastened to a bar of torsional stiffness k_T. If the body is rotated through an angle θ and released, torsional vibration of the body results.

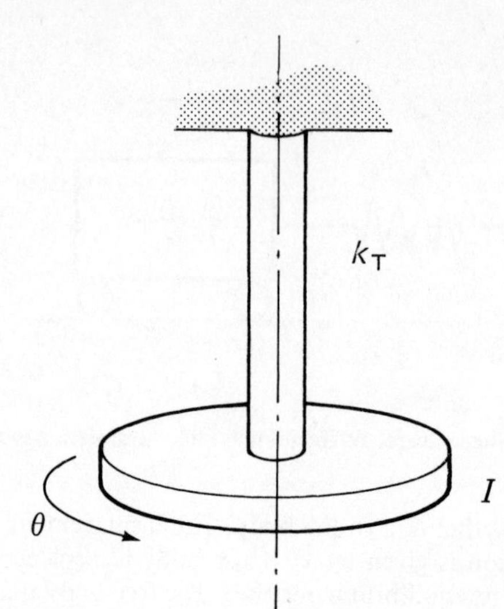

Fig. 2.3 – Single degree of freedom model – torsional vibration.

For a general displacement θ the FBDs are as given in Figs. 2.4(a) and (b). Hence the equation of motion is

$$I\ddot{\theta} = -k_T\theta \qquad \text{or} \qquad \ddot{\theta} + \frac{k_T}{I}\theta = 0$$

That is, the motion is simple harmonic with frequency $1/2\pi.\sqrt{k_T/I}$ Hz.

k_T, the torsional stiffness of the shaft, is equal to the applied torque divided by the angle of twist.

Hence $\quad k_T = \dfrac{GJ}{l}$,

where G = modulus of rigidity for shaft material,
J = second moment of area about the axis of rotation, and
l = length of shaft.

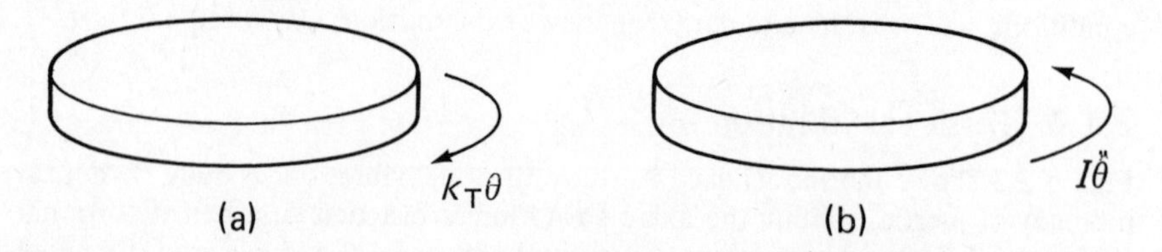

Fig. 2.4(a) – Applied torque. (b) Effective torque.

2.1.3 Energy methods for analysis

For undamped free vibration the total energy in a vibrating system is constant throughout the cycle. Therefore the maximum potential energy V_{max} is equal to the maximum kinetic energy T_{max} although these maxima occur at different times during the cycle of vibration. Furthermore, since the total energy is constant,

$$T + V = \text{constant},$$

and thus

$$\frac{d}{dt}(T + V) = 0.$$

Applying this method to the case of translational vibration above, when the body is a distance x from the equilibrium position,

$$V = \tfrac{1}{2}kx^2$$

and $$T = \tfrac{1}{2}m\dot{x}^2.$$

Thus $$\frac{d}{dt}\left(\tfrac{1}{2}m\dot{x}^2 + \tfrac{1}{2}kx^2\right) = 0$$

that is $$m\ddot{x}\,\dot{x} + k\dot{x}\,x = 0$$

or $$\ddot{x} + \frac{k}{m}x = 0, \text{ as before.}$$

Alternatively, assuming simple harmonic motion

$$x = X \cos \omega t,$$

the maximum $V = \tfrac{1}{2}kX^2$,

and maximum $T = \tfrac{1}{2}m(X\omega)^2$.

Therefore $\tfrac{1}{2}kX^2 = \tfrac{1}{2}mX^2\omega^2$, since $V_{max} = T_{max}$.

That is, $$\omega = \sqrt{\frac{k}{m}} \text{ rad/s.}$$

Energy methods can also be used in the analysis of the vibration of continuous systems such as beams. It has been shown by Rayleigh that the lowest natural frequency of such systems can be fairly accurately found by assuming any reasonable deflection curve for the vibrating shape of the beam; this is necessary for the evaluation of the kinetic and potential energies. In this way the continuous system is modelled as a single degree of freedom system, because once one coordinate of beam vibration is known, the complete beam shape during vibration is revealed. Naturally the assumed deflection curve must be compatible with the end conditions of the system, and since any deviation from the true mode shape puts additional constraints on the system, the frequency determined by Rayleigh's method is never less than the exact frequency. Generally, however, the difference is only a few per cent. The frequency of vibration is found by considering the conservation of energy in the system; the natural frequency is determined by dividing the expression for potential energy in the system by the expression for kinetic energy.

2.1.3.1 The vibration of systems with heavy springs

The mass of the spring element can have a considerable effect on the frequency of vibration of those structures in which heavy springs are used.

Consider the translational system shown in Fig. 2.5, where a rigid body of mass M is connected to a fixed frame by a spring of mass m, length l, and stiffness k. The body moves in the x direction only. If the dynamic deflected shape of the spring is assumed to be the same as the static shape, the velocity of the spring element is $\dot{y} = (y/l)\,\dot{x}$, and the mass of the element is $(m/l)\mathrm{d}y$.

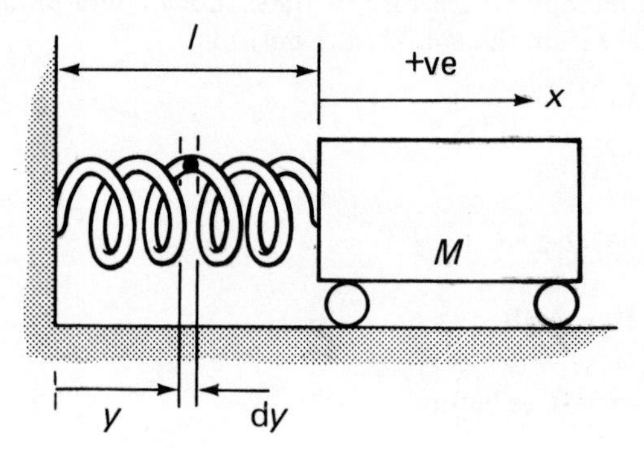

Fig. 2.5 – Single degree of freedom system with heavy spring.

Thus
$$T = \tfrac{1}{2}M\dot{x}^2 + \int_0^l \tfrac{1}{2}\left(\frac{m}{l}\right)\left[\frac{y}{l}\,\dot{x}\right]^2 \mathrm{d}y$$
$$= \tfrac{1}{2}\left(M + \frac{m}{3}\right)\dot{x}^2$$

and
$$V = \tfrac{1}{2}kx^2.$$

Assuming simple harmonic motion and putting $T_{\max} = V_{\max}$ gives the frequency of free vibration as

$$f = \frac{1}{2\pi}\sqrt{\frac{k}{M + \dfrac{m}{3}}}\ \text{Hz}.$$

That is, if the system is to be modelled with a massless spring, one third of the actual spring mass must be added to the mass of the body in the frequency calculation.

2.1.3.2 Transverse vibration of beams

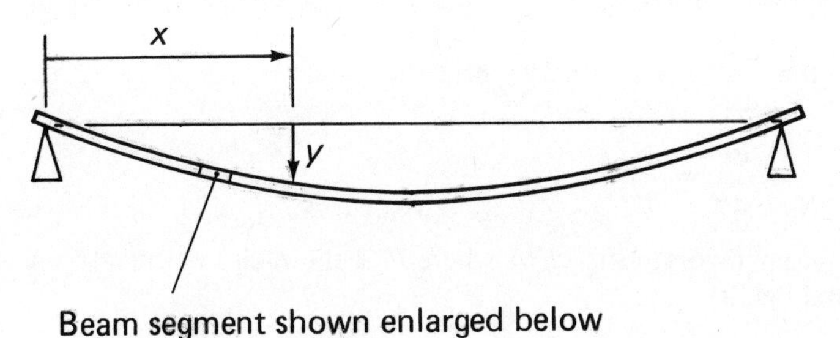

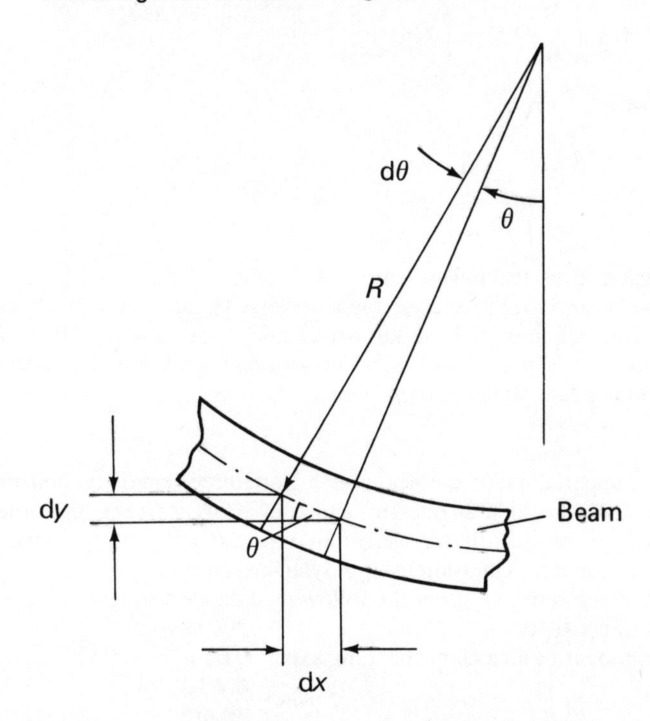

Fig. 2.6 – Beam deflection.

For the beam shown in Fig. 2.6, if m is the mass/length and y is the amplitude of the assumed deflection curve,

$$T_{max} = \tfrac{1}{2}\int \dot{y}^2_{max}\,\mathrm{d}m = \tfrac{1}{2}\,\omega^2 \int y^2\,\mathrm{d}m$$

where ω is the natural circular frequency of the beam.

The strain energy of the beam is the work done on the beam which is stored as elastic energy. If the bending moment is M and the slope of the elastic curve is θ,

$$V = \tfrac{1}{2}\int M\,\mathrm{d}\theta .$$

Usually the deflection of beams is small so that the following relationships can be assumed to hold:

$$\theta = \frac{dy}{dx} \quad \text{and} \quad R\,d\theta = dx,$$

thus
$$\frac{1}{R} = \frac{d\theta}{dx} = \frac{d^2y}{dx^2}.$$

From beam theory, $M/I = E/R$ where R is the radius of curvature and EI is the flexural rigidity.

Thus
$$V = \tfrac{1}{2}\int \frac{M}{R}\,dx = \tfrac{1}{2}\int EI\left(\frac{d^2y}{dx^2}\right)^2 dx.$$

Since $T_{max} = V_{max}$;

$$\omega^2 = \frac{\int EI\left(\dfrac{d^2y}{dx^2}\right)^2 dx}{\int y^2\,dm}.$$

This expression gives the lowest natural frequency of transverse vibration of a beam. It can be seen that to analyse the transverse vibration of a particular beam by this method requires y to be known as a function of x. For this the static deflected shape or a part sinusoid can be assumed, provided the shape is compatible with the beam boundary conditions.

Example 1

Part of an industrial plant incorporates a horizontal length of uniform pipe, which is rigidly embedded at one end and is effectively free at the other. Considering the pipe as a cantilever, derive an expression for the frequency of the first mode of transverse vibration using Rayleigh's method.

Calculate this frequency, given the following data for the pipe:

Modulus of elasticity	200 GN/m^2
Second moment of area about bending axis	0.02 m^4
Mass	6 x 10^4 kg
Length	30 m
Outside diameter	1 m

For a cantilever,

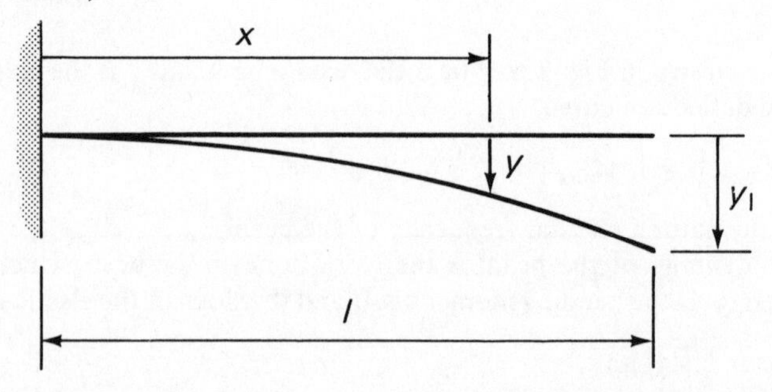

assume $y = y_l \left(1 - \cos \dfrac{\pi x}{2l}\right)$.

This is compatible with zero deflection and slope when $x = 0$, and zero shear force and bending moment when $x = l$.

Thus $$\frac{\mathrm{d}^2 y}{\mathrm{d}x^2} = y_l \left(\frac{\pi}{2l}\right)^2 \cos \frac{\pi x}{2l}.$$

Now $$\int_0^l EI \left(\frac{\mathrm{d}^2 y}{\mathrm{d}x^2}\right)^2 \mathrm{d}x = EI \int_0^l y_l^2 \left(\frac{\pi}{2l}\right)^4 \cos^2 \frac{\pi x}{2l} \, \mathrm{d}x$$

$$= EI.\, y_l^2 \left(\frac{\pi}{2l}\right)^4 . \frac{l}{2},$$

and $$\int_0^l y^2 \, \mathrm{d}m = \int_0^l y_l^2 \left(1 - \cos \frac{\pi x}{2l}\right)^2 \frac{m}{l} \, \mathrm{d}x$$

$$= y_l^2 \, m \left(\frac{3}{2} - \frac{4}{\pi}\right).$$

Hence, assuming the structure to be conservative, that is the total energy remains constant throughout the vibration cycle,

$$\omega^2 = \frac{EI.y_l^2 \left(\dfrac{\pi}{2l}\right)^4 . \dfrac{l}{2}}{y_l^2 \, m \left(\dfrac{3}{2} - \dfrac{4}{\pi}\right)}$$

$$= \frac{EI}{ml^3} \cdot 13.4$$

Thus $$\omega = 3.66 \sqrt{\frac{EI}{ml^3}},$$

and $$f = \frac{3.66}{2\pi} \sqrt{\frac{EI}{ml^3}} \text{ Hz}.$$

In this case $\dfrac{EI}{ml^3} = \dfrac{200 \,.\, 10^9 \,.\, 0.02}{6 \,.\, 10^4 \,.\, 30^3} \Big/ \mathrm{s}^2$.

Hence $\omega = 5.75$ rad/s

and $f = 0.92$ Hz.

2.1.4 The stability of vibrating structures

If a structure is to vibrate about an equilibrium position, it must be stable about that position. That is, if the structure is disturbed when in an equilibrium position, the elastic forces must be such that the structure vibrates about the equilibrium position. Thus the expression for ω^2 must be positive if a real value of the frequency of vibration about the equilibrium position is to exist, and hence the potential energy of a stable structure must also be positive.

The principle of minimum potential energy can be used to test the stability of structures which are conservative. Thus a structure will be stable at an equilibrium position if the potential energy of the structure is a minimum at that position. This requires that

$$\frac{dV}{dq} = 0, \quad \text{and} \quad \frac{d^2 V}{dq^2} > 0,$$

where q is an independent or generalised coordinate. Hence the necessary conditions for vibration to take place are found, and the position about which the vibration occurs is determined.

Example 2

A uniform building of height $2h$ and mass m has a rectangular base $a \times b$ which rests on an elastic soil. The stiffness of the soil, k is expressed as the force per unit area required to produce unit deflection.

Find the lowest frequency of free low amplitude swaying oscillation of the building.

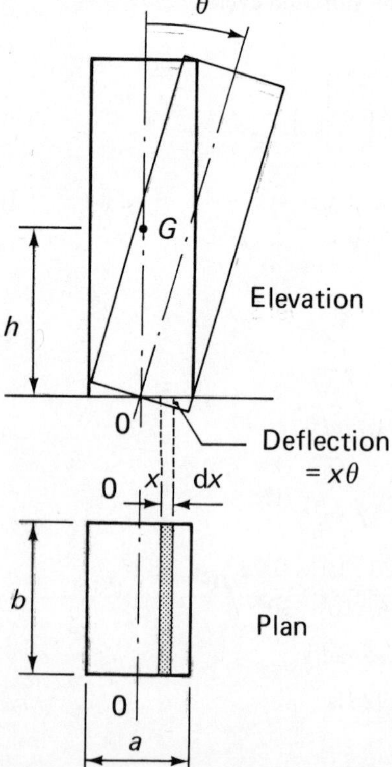

The lowest frequency of oscillation about the axis 0–0 through the base of the building, is when the oscillation occurs about the shortest side, length a.

I_0 is the mass moment of inertia of the building about axis 0–0.

The FBDs are:

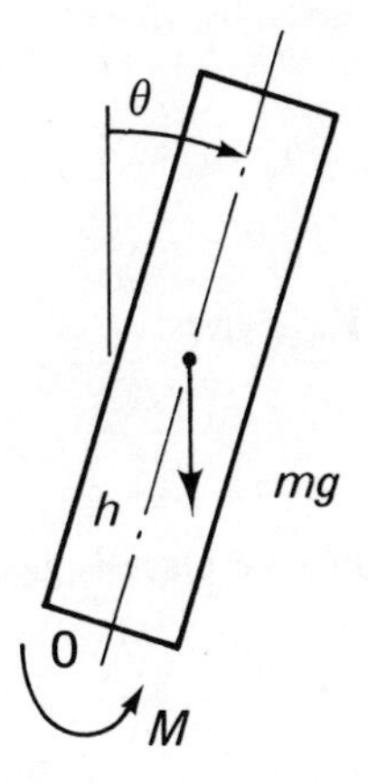

$I_0\ddot{\theta}$

0

and the equation of motion for small θ is given by:

$$I_0\ddot{\theta} = mgh\theta - M,$$

where M is the restoring moment from the elastic soil.

For the soil, k = force/(area × deflection), so considering an element of the base as shown, the force on element = $k.b\,\mathrm{d}x.x\theta$, and the moment of this force about axis 0–0 = $kb\,\mathrm{d}x.x\theta.x$.

Thus the total restoring moment M, assuming the soil acts similarly in tension and compression is:

$$M = 2\int_0^{a/2} kbx^2\theta\,\mathrm{d}x$$

$$= 2kb\theta\,\frac{(a/2)^3}{3} = \frac{ka^3b}{12}\theta.$$

Thus the equation of motion becomes:

$$I_0\ddot{\theta} + \left(\frac{ka^3b}{12} - mgh\right)\theta = 0.$$

Motion is therefore simple harmonic, with frequency

$$f = \frac{1}{2\pi}\sqrt{\frac{ka^3b/12 - mgh}{I_0}}\ \text{Hz}.$$

An alternative solution can be obtained by considering the energy in the system. In this case,

$$T = \tfrac{1}{2}.I_0.\dot{\theta}^2,$$

and $$V = \tfrac{1}{2}.2.\int_0^{a/2} kb\,\mathrm{d}x.x\theta.x\theta - \frac{mgh\theta^2}{2},$$

where the loss in potential energy of building weight is given by $mgh\,(1 - \cos\theta) \triangleq mgh\,\theta^2/2$, since $\cos\theta \triangleq 1 - \theta^2/2$.

Thus $$V = \left(\frac{ka^3b}{24} - \frac{mgh}{2}\right)\theta^2.$$

Assuming simple harmonic motion, and putting $T_{\max} = V_{\max}$ gives:

$$\omega^2 = \frac{ka^3b/12 - mgh}{I_0}.$$

The equilibrium position about which oscillation will take place is given by $\mathrm{d}V/\mathrm{d}\theta = 0$, that is:

$$\left(\frac{ka^3b}{24} - \frac{mgh}{2}\right)2\theta = 0.$$

Thus $\theta = 0$ is the equilibrium position.

Also for stable oscillation, $\dfrac{\mathrm{d}^2V}{\mathrm{d}\theta^2} > 0$,

thus $$\left(\frac{ka^3b}{24} - \frac{mgh}{2}\right)2 > 0.$$

That is $ka^3b > 12\,mgh$.

This expression gives the minimum value of k, the soil stiffness, for stable oscillation of a particular building to occur. If k is less than $12mgh/a^3b$ the building will fall over when disturbed. This can also be seen by considering the expression for the frequency of oscillation; if k is equal to $12mgh/a^3b$, $f = 0$.

2.2 FREE DAMPED VIBRATION

All structures dissipate energy when they vibrate. The energy dissipated is often very small so that an undamped analysis is sometimes realistic; but when the damping is significant its effect must be included in the analysis, particularly when the amplitude of vibration is required. The damping which occurs in structures is due to frictional effects such as that occurring at the connection between elements, or internal friction in the structural members. It is often difficult to model damping exactly because many mechanisms may be operating in a structure. However, each type of damping can be analysed, and since in many structures one form of damping predominates, a reasonably accurate analysis is usually possible. The most common types of damping are **viscous, dry friction** and **hysteretic**. Hysteretic damping arises in structural elements due to hysteresis losses in the material.

2.2.1 Vibration with viscous damping

Viscous damping is a common form of damping where the damping force is proportional to the first power of the velocity across the damper. The damping force always opposes the motion so that it is a continuous linear function of the velocity. Because viscous type damping can be expressed in a simple mathematical way, other more complex types of damping are often expressed as an equivalent viscous damping in the analysis.

Consider the single degree of freedom system shown in Fig. 2.7. The viscous

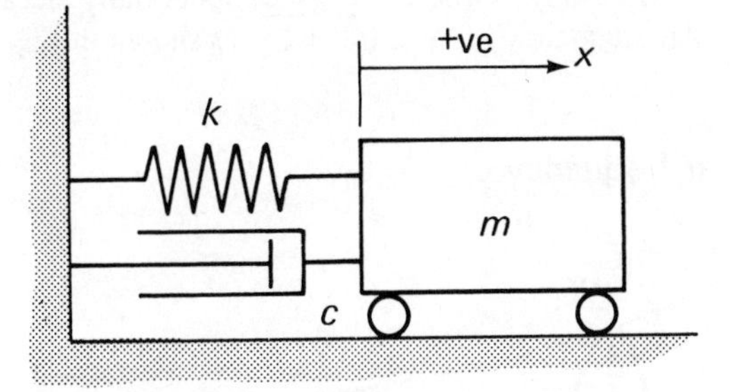

Fig. 2.7 – Single degree of freedom model with viscous damping.

damper has a coefficient c, which means that the damping force opposing the motion of the body is $c\dot{x}$. The equation of motion is therefore:

$$m\ddot{x} + c\dot{x} + kx = 0$$

This equation can be solved by assuming a solution of the form $x = Xe^{st}$. Substituting this solution into the equation of motion gives

$$(ms^2 + cs + k)\,Xe^{st} = 0,$$

that is $\quad ms^2 + cs + k = 0.$

Hence $$s_{1,2} = -\frac{c}{2m} \pm \frac{\sqrt{c^2 - 4mk}}{2m}.$$

Thus $$x = X_{\mathrm{I}}e^{s_1 t} + X_{\mathrm{II}}e^{s_2 t}$$

X_{I} and X_{II} are arbitrary constants found from the initial conditions. The system response evidently depends on whether c is positive or negative and on whether c^2 is greater than, equal to, or less than $4mk$.

The critical damping coefficient c_c is chosen to make the radical zero. Thus

$$c_c = 2\sqrt{mk}.$$

The actual damping in a system can be specified in terms of c_c by introducing the damping ratio ζ where

$$\zeta = \frac{c}{c_c}.$$

Thus $$s_{1,2} = (-\zeta \pm \sqrt{\zeta^2 - 1})\,\omega.$$

In structures the damping is usually very low so that $\zeta < 1$, that is the damping is less than critical. In this case

$$s_{1,2} = -\zeta\omega \pm j\sqrt{1-\zeta^2}\,\omega, \qquad \text{where } j = \sqrt{-1},$$

so $$x = e^{-\zeta\omega t}\,[X_{\mathrm{I}} e^{j\sqrt{1-\zeta^2}\,\omega t} + X_{\mathrm{II}} e^{-j\sqrt{1-\zeta^2}\,\omega t}]$$

and $$x = Xe^{-\zeta\omega t} \sin(\sqrt{1-\zeta^2}\,\omega t + \phi).$$

The motion of the body is therefore an exponentially decaying harmonic motion with circular frequency $\omega_v = \omega\sqrt{1-\zeta^2}$, as shown in Fig. 2.8.

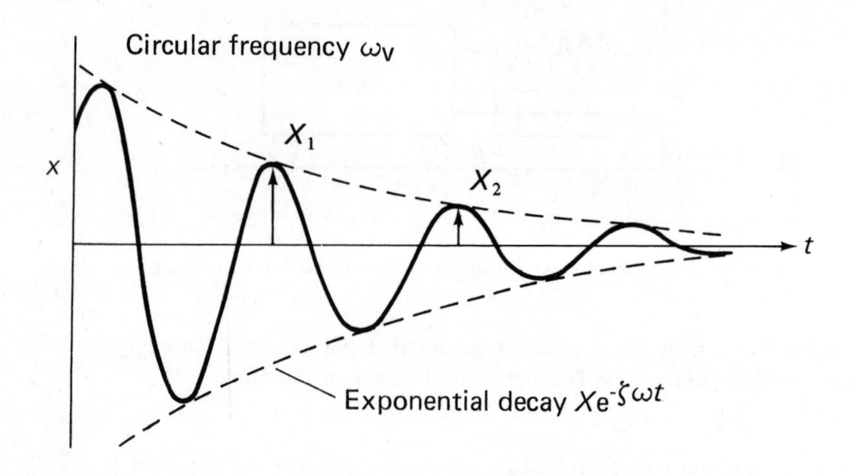

Fig. 2.8 – Vibration decay of system with viscous damping.

Since ζ is very small in most structures, ω_v is usually almost identical to ω.

The solution for $\zeta = 1$ and $\zeta > 1$ need not be considered here because these conditions will rarely, if ever, arise in structures. However in these cases the displacement decays with time and without oscillation.

Since $\omega_v \doteq \omega$, ζ cannot be determined with any accuracy from the equation $\omega_v = \omega\sqrt{1-\zeta^2}$. However ζ can be found from the rate of decay of the oscillations, as follows.

The **logarithmic decrement**, Λ is the natural logarithm of the ratio of any two successive amplitudes in the same direction, and thus from Fig. 2.8,

$$\Lambda = \ln\left(\frac{X_1}{X_2}\right).$$

Since $\quad x = Xe^{-\zeta\omega t}\sin(\omega_v t + \phi)$,

if $\quad X_1 = Xe^{-\zeta\omega t},\ X_2 = Xe^{-\zeta\omega(t+\tau_v)}$,

where τ_v is the period of the damped oscillation.

Thus $$\Lambda = \ln\frac{Xe^{-\zeta\omega t}}{Xe^{-\zeta\omega(t+\tau_v)}} = \zeta\omega\tau_v.$$

Since $$\tau_v = \frac{2\pi}{\omega_v} = \frac{2\pi}{\omega\sqrt{1-\zeta^2}},$$

$$\Lambda = \frac{2\pi\zeta}{\sqrt{1-\zeta^2}} = \ln\left(\frac{X_1}{X_2}\right).$$

Example 3

Consider the transverse vibration of a bridge structure. For the fundamental frequency it can be considered as a single degree of freedom system. The bridge is deflected at mid-span (by winching the bridge down) and suddenly released. After the initial disturbance the vibration was found to decay exponentially from an amplitude of 10 mm to 5.8 mm in 3 cycles with a frequency of 1.62 Hz. The test was repeated with a vehicle of mass 40 000 kg at mid-span, and the frequency of free vibration was measured to be 1.54 Hz.

Find the effective mass, the effective stiffness, and the damping ratio of the structure.

Let m be the effective mass and k the effective stiffness.

$$f_1 = 1.62 = \frac{1}{2\pi}\sqrt{\frac{k}{m}},$$

and $$f_2 = 1.54 = \frac{1}{2\pi}\sqrt{\frac{k}{m + 40\,.\,10^3}}.$$

Thus $$\left(\frac{1.62}{1.54}\right)^2 = \frac{m + 40\,.\,10^3}{m},$$

hence $$m = 375\,.\,10^3\ \text{kg}.$$

Since $$k = (2\pi f_1)^2 m,$$

$$k = 38\,850\ \text{kN/m}.$$

Now $$\Lambda = \ln\frac{X_1}{X_2} = \tfrac{1}{3}\,.\,\ln\frac{X_1}{X_4} = \tfrac{1}{3}\,.\,\ln\left(\frac{10}{5.8}\right)$$

$$= 0.182.$$

Thus $$\Lambda = \frac{2\pi\zeta}{\sqrt{1-\zeta^2}} = 0.182$$

and hence $$\zeta = 0.029.$$

(This compares with a value of about 0.002 for cast iron material. The additional damping originates in the joints of the structure.)

2.2.2 Vibration with coulomb (dry friction) damping

In many structures, steady friction forces occur when relative motion takes place between adjacent members. These forces are independent of amplitude and frequency, they always oppose the motion, and their magnitude may, to a first approximation, be considered constant. Dry friction damping in a single degree of freedom system can be modelled as shown in Fig. 2.9.

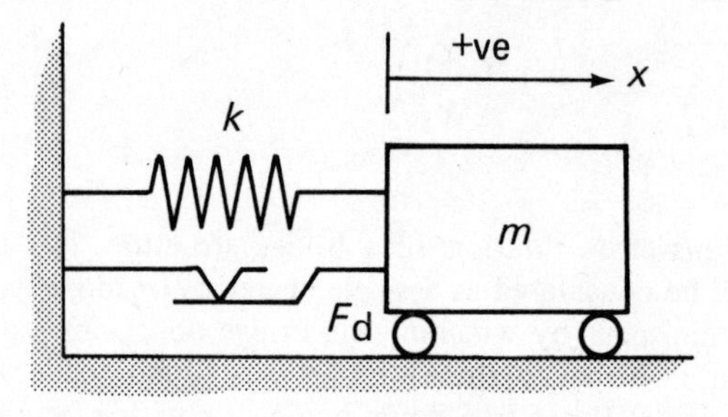

Fig. 2.9 – Single degree of freedom model with coulomb damping.

The friction force F_d always opposes the motion, so that if the body is displaced a distance x_0 to the right and released from rest we have, for motion from right to left only,

$$m\ddot{x} = F_d - kx$$

or $$m\ddot{x} + kx = F_d.$$

The solution to this equation of motion is

$$x = A \sin \omega t + B \cos \omega t + \frac{F_d}{k}$$

where $$\omega = \sqrt{\frac{k}{m}} \text{ rad/s.}$$

The initial conditions were $x = x_0$ at $t = 0$, and $\dot{x} = 0$ at $t = 0$. Thus

$$A = 0 \quad \text{and} \quad B = x_0 - \frac{F_d}{k}.$$

Hence $$x = \left(x_0 - \frac{F_d}{k}\right) \cos \omega t + \frac{F_d}{k}.$$

At the end of the half cycle right to left, $\omega t = \pi$ and

$$x_{(t=\pi/\omega)} = -x_0 + \frac{2F_d}{k}.$$

That is, a reduction in amplitude of $2F_d/k$ per half cycle occurs.

From symmetry, for motion from left to right when the friction force acts in the opposite direction to the above, the initial displacement is $(x_0 - 2F_d/k)$ and the final displacement is therefore $(x_0 - 4F_d/k)$, i.e. the reduction in amplitude is $4F_d/k$ per cycle. This oscillation continues until the amplitude of the motion is so small that the spring force is unable to overcome the friction force F_d. This can happen whenever the amplitude is $\leqslant \pm (F_d/k)$. The motion is therefore sinusoidal for each half cycle, with successive half cycles centred on points distant $+F_d/k$ and $-F_d/k$ from the origin. The oscillation ceases with $|x| \leqslant F_d/k$, and the frequency of the oscillation is $1/2\pi.\sqrt{k/m}$ Hz.

Example 4

Part of a structure can be modelled as a torsional system comprising a bar of stiffness 10 kNm/rad and a beam of moment inertia about the axis of rotation of 50 kgm^2. The bottom guide imposes a friction torque of 10 Nm.

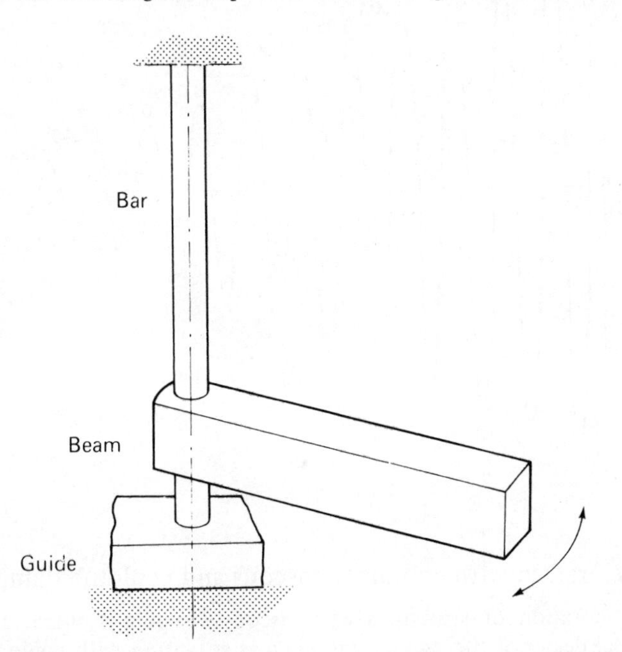

If the beam is displaced through 0.05 rad from its equilibrium position, and released, find the frequency of the oscillation, the number of cycles executed before the beam motion ceases, and the position of the beam when this happens.

$$\omega = \sqrt{\frac{k_T}{I}} = \sqrt{\frac{10.10^3}{50}} = 14.14 \text{ rad/s}.$$

Thus $$f = \frac{14.14}{2\pi} = 2.25 \text{ Hz}.$$

$$\text{Loss in amplitude/cycle} = \frac{4F_d}{k} = \frac{4.10}{10^4} \text{ rad}$$

$$= 0.004 \text{ rad.}$$

Number of cycles for motion to cease

$$= \frac{0.05}{0.004} = 12\tfrac{1}{2}.$$

The beam is in the initial (equilibrium) position when motion ceases. The motion is shown in the figure below.

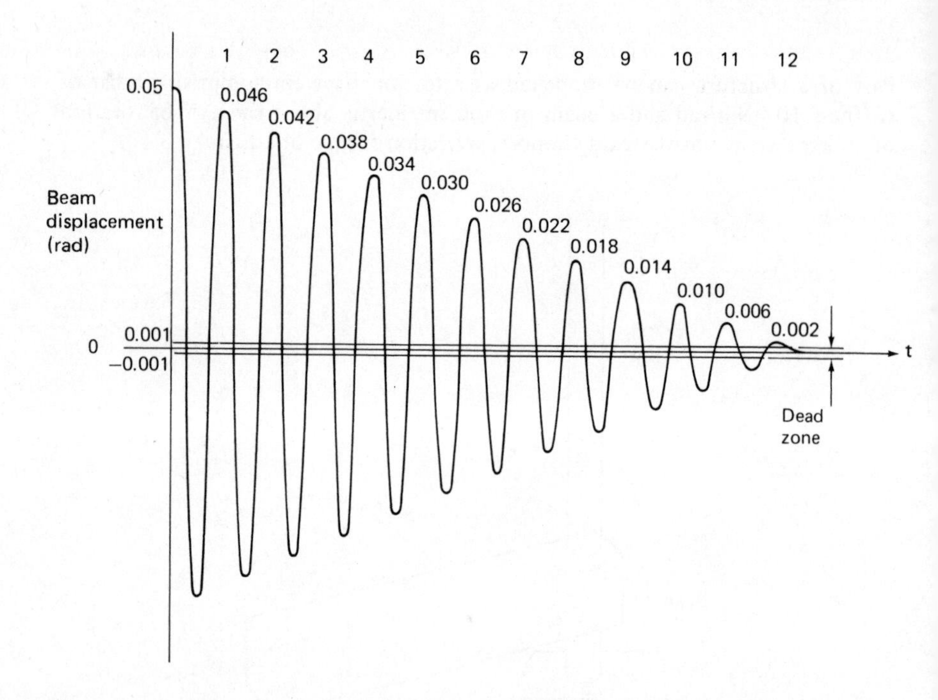

2.2.3 Vibration with combined viscous and coulomb damping

The free vibration of structures with viscous damping is characterised by an exponential decay of the oscillation, whereas structures with coulomb damping possess a linear decay of oscillation. Many real structures have both forms of damping, so that their vibration decay is a combination of exponential and linear functions.

The two damping actions are sometimes amplitude dependent, so that initially the decay is exponential, say, and only towards the end of the oscillation does the coulomb effect show. In the analyses of these cases the coulomb effect can easily be separated from the total damping to leave the viscous damping alone. The exponential decay with viscous damping can be checked by plotting the amplitudes on logarithmic–linear axes when the decay should be seen to be linear.

If the coulomb and viscous effects cannot be separated in this way, a mixture of linear and exponential decay functions have to be found by trial and error in order to conform with the experimental data.

2.2.4 Vibration with hysteretic damping

Experiments on the damping that occurs in solid materials and structures which have been subjected to cyclic stressing have shown the damping force to be independent of frequency. This internal, or material, damping is referred to as hysteretic damping. Since the viscous damping force $c\dot{x}$ is dependent on the frequency of oscillation, it is not a suitable way of modelling the internal damping of solids and structures. The analysis of structures with this form of damping therefore requires the damping force $c\dot{x}$ to be divided by the frequency of oscillation ω. Thus the equation of motion becomes $m\ddot{x} + (c/\omega)\,\dot{x} + kx = 0$.

However, it has been observed from experiments carried out on many materials and structures that under harmonic forcing the stress leads the strain by a constant angle, α.

Thus for an harmonic strain, $\epsilon = \epsilon_0 \sin \nu t$,
the induced stress is $\sigma = \sigma_0 \sin (\nu t + \alpha)$.

Hence
$$\begin{aligned}\sigma &= \sigma_0 \cos \alpha \sin \nu t + \sigma_0 \sin \alpha \cos \nu t \\ &= \sigma_0 \cos \alpha \sin \nu t + \sigma_0 \sin \alpha \sin \left(\nu t + \frac{\pi}{2}\right).\end{aligned}$$

The first component of stress is in-phase with the strain ϵ, whilst the second component is in quadrature with ϵ and $\pi/2$ ahead. Putting $j = \sqrt{-1}$,

$$\sigma = \sigma_0 \cos \alpha \sin \nu t + j\, \sigma_0 \sin \alpha \sin \nu t.$$

Hence a complex modulus E^* can be formulated where;

$$\begin{aligned}E^* = \frac{\sigma}{\epsilon} &= \frac{\sigma_0}{\epsilon_0} \cos \alpha + j\, \frac{\sigma_0}{\epsilon_0} \sin \alpha \\ &= E' + jE'',\end{aligned}$$

where E' is the in-phase or storage modulus, and E'' is the quadrature or loss modulus.

The loss factor η, which is a measure of the hysteretic damping in a structure, is equal to E''/E'.

It is not usually possible to separate the stiffness of a structure from its hysteretic damping, so that in a mathematical model these quantities have to be considered together. The complex stiffness k^* is given by $k^* = k(1 + j\eta)$, where k is the static stiffness and η the hysteretic damping loss factor.

The equation of free motion for a single degree of freedom structure with hysteretic damping is therefore $m\ddot{x} + k^*x = 0$.

2.2.5 Energy dissipated by damping

The energy dissipated per cycle by the viscous damping force in a single degree of freedom vibrating system is approximately

$$4\int_0^X c\dot{x}\,\mathrm{d}x,$$

if $x = X \sin \omega t$ is assumed for the complete cycle. The energy dissipated is therefore

$$4\int_0^{\pi/2} cX^2\,\omega^2 \cos \omega t\,\mathrm{d}t$$

$$= \pi c\omega X^2 .$$

The energy dissipated per cycle by coulomb damping is $4F_\mathrm{d}X$ approximately. Thus an equivalent viscous damping coefficient, c_d for coulomb damping can be deduced where

$$\pi c_\mathrm{d}\omega X^2 = 4F_\mathrm{d}X,$$

that is,
$$c_\mathrm{d} = \frac{4F_\mathrm{d}}{\pi\omega X}.$$

The energy dissipated per cycle by a force F acting on a structure with hysteretic damping is $\int F\mathrm{d}x$ where $F = k^*x = k(1 + j\eta)x$, and x is the displacement.

For harmonic motion $x = X \sin \omega t$,

so
$$F = kX \sin \omega t + j\eta kX \sin \omega t$$
$$= kX \sin \omega t + \eta kX \cos \omega t.$$

Now $\sin \omega t = \dfrac{x}{X}$, therefore $\cos \omega t = \dfrac{\sqrt{X^2 - x^2}}{X}$.

Thus
$$F = kx \pm \eta k\sqrt{X^2 - x^2}.$$

This is the equation of an ellipse as shown in Fig. 2.10. The energy dissipated is given by the area enclosed by the ellipse.

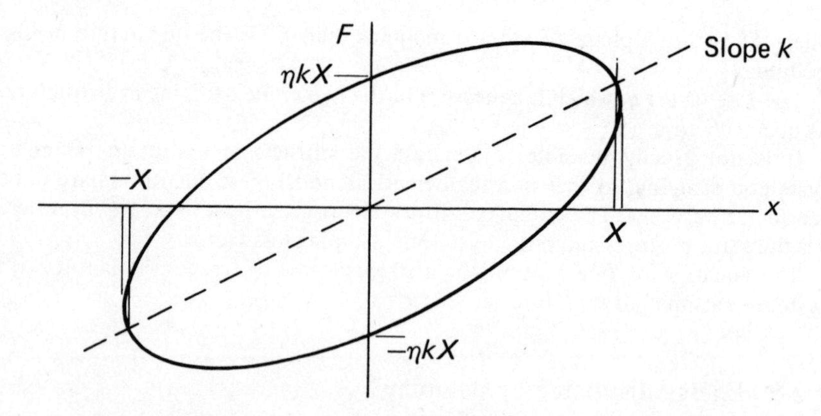

Fig. 2.10 – Elliptical force-displacement relationship for a structure with hysteretic damping.

Hence $$\int F\,\mathrm{d}x = \int_0^X (kx \pm \eta k\sqrt{X^2 - x^2})\,\mathrm{d}x$$

$$= \pi X^2 \eta k.$$

An equivalent viscous damping coefficient c_H is given by

$$\pi c_H \omega X^2 = \pi \eta k X^2,$$

that is $$c_H = \frac{\eta k}{\omega} \quad \text{or} \quad c_H = \frac{c}{\omega}.$$

2.3 FORCED VIBRATION

Many structures are exposed to forced excitation such as can arise from a shaking foundation or a steady cross wind. Such excitation is usually periodic and, since any periodic function can be expressed as the sum of a number of harmonic components, it is relevant to consider the vibration of structures subjected to harmonic excitation.

2.3.1 Response of a viscous damped structure to a simple harmonic exciting force with constant amplitude

In the system shown in Fig. 2.11, the body of mass m is connected by a spring and viscous damper to a fixed support, whilst an harmonic force $F \sin \nu t$ acts upon it in the line of motion. The equation of motion is

$$m\ddot{x} + c\dot{x} + kx = F \sin \nu t.$$

The solution to this equation comprises a complementary function and a particular solution. The complementary function represents the transient initial vibra-

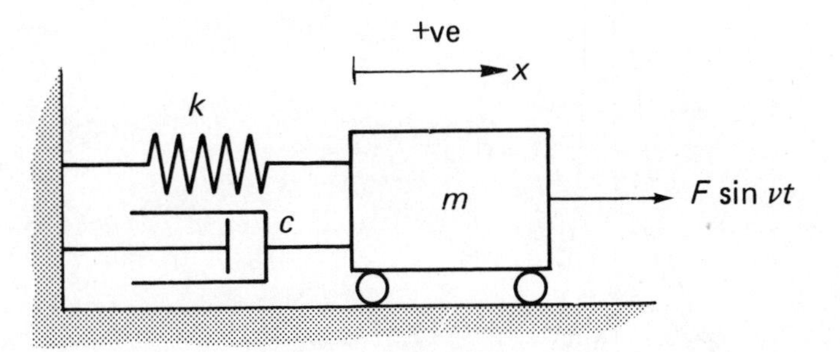

Fig. 2.11 – Single degree of freedom model of a forced structure with viscous damping.

tion which dies away, (section 2.2.1) whereas the particular solution represents the sustained motion. For the sustained motion a solution $x = X \sin (\nu t - \phi)$ can be assumed, because this represents simple harmonic motion at the frequency of the exciting force, with a displacement vector which lags the force vector by the phase angle, ϕ, that is, the motion occurs after the application of the force.

Substituting this value of x into the equation of motion and solving for x gives

$$x = \frac{F}{\sqrt{(k - m\nu^2)^2 + (c\nu)^2}} \sin(\nu t - \phi),$$

where $\phi = \tan^{-1} \dfrac{c\nu}{k - m\nu^2}$.

If $\omega = \sqrt{\dfrac{k}{m}}$ rad/s and $X_s = \dfrac{F}{k}$ is substituted, then

$$\frac{X}{X_s} = \frac{1}{\sqrt{\left[1 - \left(\frac{\nu}{\omega}\right)^2\right]^2 + \left[2\zeta\frac{\nu}{\omega}\right]^2}},$$

and $\phi = \tan^{-1} \dfrac{2\zeta \dfrac{\nu}{\omega}}{1 - \left(\dfrac{\nu}{\omega}\right)^2}$.

X/X_s is known as the **dynamic magnification factor**, because X_s is the static deflection of the system under a steady force F. X/X_s and ϕ are plotted as functions of the frequency ratio ν/ω in Figs. 2.12 and 2.13.

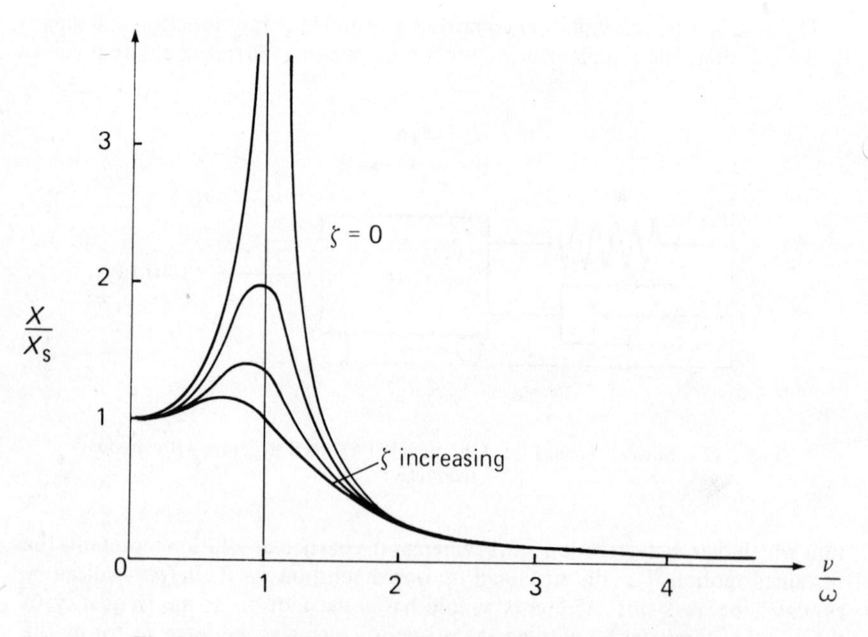

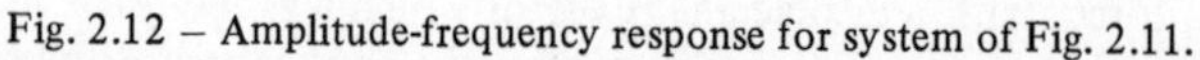
Fig. 2.12 – Amplitude-frequency response for system of Fig. 2.11.

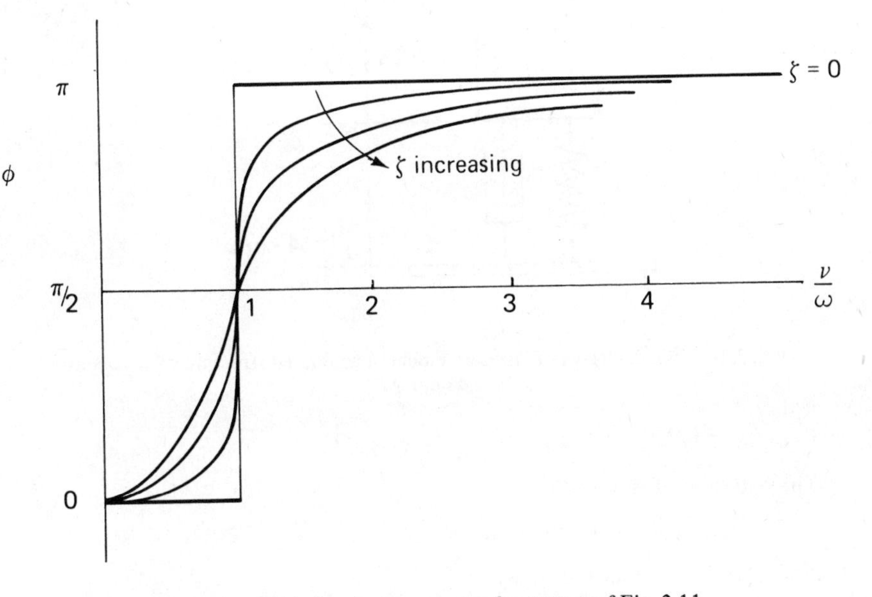

Fig. 2.13 – Phase-frequency response for system of Fig. 2.11.

The importance of vibration in structures arises mainly from the large values of X/X_s experienced in practice when ν/ω has a value near unity: this means that a small harmonic force can produce a large amplitude of vibration. The phenomenon known as **resonance** occurs when $\nu = \omega$. The maximum value of X/X_s actually occurs at values of ν/ω less than unity. For small values of ζ, $(X/X_s)_{max} = 1/2\zeta$. $1/2\zeta$ is a measure of the damping in a structure and is referred to as the Q factor.

Note. An alternative solution to the equation of motion can be obtained by putting $F \sin \nu t = \text{Im. } Fe^{j\nu t}$.

Then $m\ddot{x} + c\dot{x} + kx = Fe^{j\nu t}$, and a solution $x = Xe^{j\nu t}$ can be assumed.

Thus $(k - m\nu^2)X + jc\nu X = F$,

or
$$X = \frac{F}{(k - m\nu^2) + jc\nu}.$$

Hence
$$X = \frac{F}{\sqrt{(k - m\nu^2)^2 + (c\nu)^2}}.$$

2.3.2 Response of a viscous damped structure supported on a foundation subjected to harmonic vibration

The structure is modelled by the system shown in Fig. 2.14. The foundation is subjected to harmonic vibration $A \sin \nu t$, and it is required to find the resulting absolute motion of the body, x.

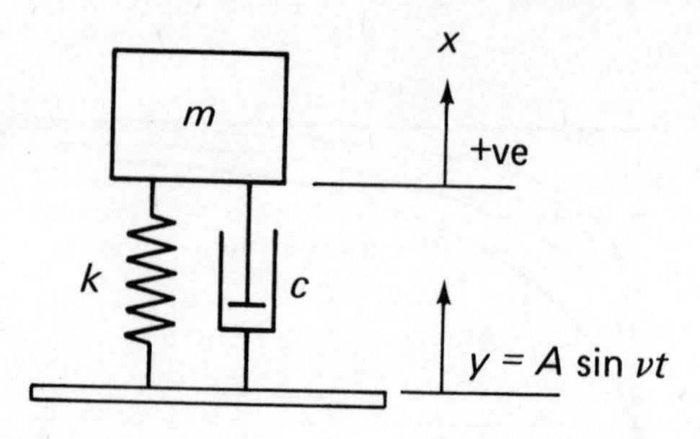

Fig. 2.14 – Single degree of freedom model of a vibrated structure with viscous damping.

The equation of motion is

$$m\ddot{x} = c(\dot{y} - \dot{x}) + k(y - x)$$

that is,

$$m\ddot{x} + c\dot{x} + kx = c\dot{y} + ky$$

$$= cA\nu \cos \nu t + kA \sin \nu t$$

$$= A\sqrt{k^2 + (c\nu)^2} \sin(\nu t + \alpha)$$

$$\text{where } \alpha = \tan^{-1}\frac{c\nu}{k}.$$

Hence, from the previous result,

$$x = \frac{A\sqrt{k^2 + (c\nu)^2}}{\sqrt{(k - m\nu^2)^2 + (c\nu)^2}} \sin(\nu t - \phi + \alpha)$$

The **motion transmissibility** is defined as the ratio of the amplitude of the body vibration to the amplitude of the foundation vibration. Thus,

$$\text{Motion transmissibility} = \frac{X}{A}$$

$$= \frac{\sqrt{1 + \left(2\zeta\dfrac{\nu}{\omega}\right)^2}}{\sqrt{\left[1 - \left(\dfrac{\nu}{\omega}\right)^2\right]^2 + \left[2\zeta\dfrac{\nu}{\omega}\right]^2}}.$$

2.3.2.1 Vibration isolation

The dynamic forces produced by machinery are often very large. However, the force transmitted to the foundation or supporting structure can be reduced by using flexible mountings with the correct properties; alternatively a machine can be isolated from foundation vibration by using the correct flexible mountings. Such a system is shown in Fig. 2.15.

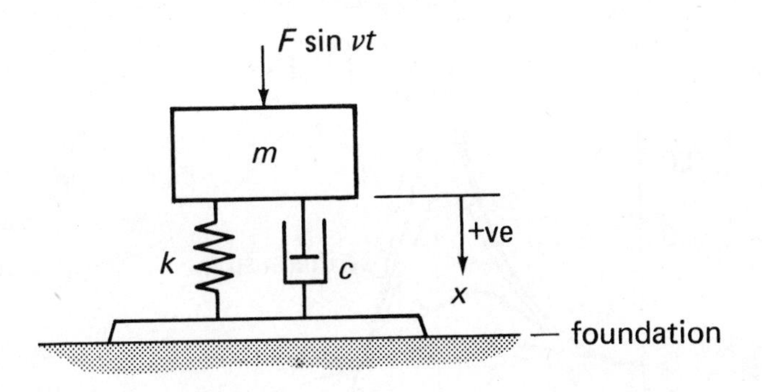

Fig. 2.15 – Single degree of freedom system with foundation.

The force transmitted to the foundation is the sum of the spring force and the damper force. Thus transmitted force = $kx + c\dot{x}$ and F_T, the amplitude of the transmitted force,

$$= \sqrt{(kX)^2 + (c\nu X)^2}\,.$$

The **force transmission ratio** or **transmissibility**, T_R is given by

$$T_R = \frac{F_T}{F} = \frac{X\sqrt{k^2 + (c\nu)^2}}{F}\,.$$

since

$$X = \frac{F/k}{\sqrt{\left[1 - \left(\frac{\nu}{\omega}\right)^2\right]^2 + \left[2\zeta\frac{\nu}{\omega}\right]^2}}$$

$$T_R = \frac{\sqrt{1 + 2\zeta\left(\frac{\nu}{\omega}\right)^2}}{\sqrt{\left[1 - \left(\frac{\nu}{\omega}\right)^2\right]^2 + \left[2\zeta\frac{\nu}{\omega}\right]^2}}$$

Thus the force and motion transmissibilities are the same.

The effect of ν/ω on T_R is shown in Fig. 2.16.

It can be seen that for good isolation $\nu/\omega \gg \sqrt{2}$, hence a low value of ω is required which implies a low stiffness, i.e. a flexible mounting: this may not always be acceptable in practice where a certain minimum stiffness is usually necessary to satisfy operating criteria.

Example 5

The vibration of the floor in a building is SHM at a frequency in the range 15–60 Hz. It is desired to install sensitive equipment in the building which must be insulated from floor vibration. The equipment is fastened to a small platform

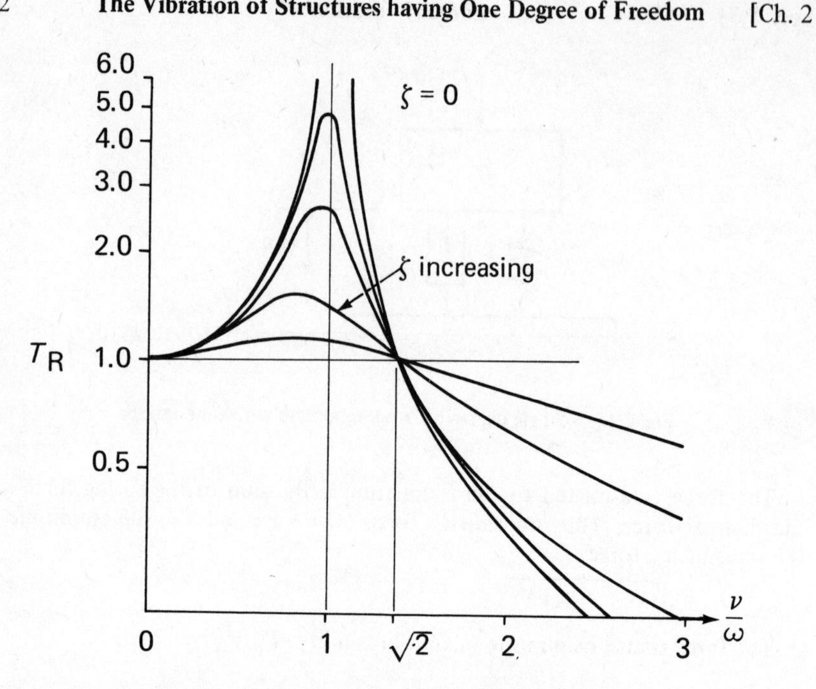

Fig. 2.16 – Transmissibility-frequency ratio response.

which is supported by three similar springs resting on the floor, each carrying an equal load. Only vertical motion occurs. The combined mass of the equipment and platform is 40 kg, and the equivalent viscous damping ratio of the suspension is 0.2.

Find the maximum value for the spring stiffness, if the amplitude of transmitted vibration is to be less than 10% of the floor vibration over the given frequency range.

$$T_R = \frac{\sqrt{1 + \left(2\zeta \dfrac{\nu}{\omega}\right)^2}}{\sqrt{\left[1 - \left(\dfrac{\nu}{\omega}\right)^2\right]^2 + \left[2\zeta \dfrac{\nu}{\omega}\right]^2}}$$

$T_R = 0.1$ with $\zeta = 0.2$ is required.

Thus
$$\left[1 - \left(\frac{\nu}{\omega}\right)^2\right]^2 + \left[0.4 \frac{\nu}{\omega}\right]^2 = 100 \left[1 + \left(0.4 \frac{\nu}{\omega}\right)^2\right]$$

i.e.
$$\left(\frac{\nu}{\omega}\right)^4 - 17.84 \left(\frac{\nu}{\omega}\right)^2 - 99 = 0.$$

Hence
$$\frac{\nu}{\omega} = 4.72.$$

When $\nu = 15.2\pi$ rad/s, $\omega = 19.97$ rad/s.

Since $\omega = \sqrt{\dfrac{k}{m}}$ and $m = 40$ kg,

total $k = 15\,935$ N/m.

i.e. stiffness of each spring $= \dfrac{15\,935}{3}$ N/m = 5.3 kN/m.

The amplitude of the transmitted vibration will be less than 10% at frequencies above 15 Hz.

2.3.3 Response of a coulomb damped structure to a simple harmonic exciting force with constant amplitude

In the system shown in Fig. 2.17 the damper relies upon dry friction.

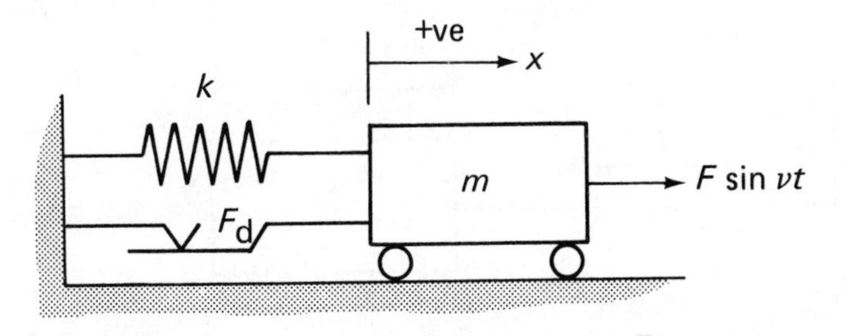

Fig. 2.17 – Single degree of freedom model of a forced structure with coulomb damping.

The equation of motion is non-linear because the friction force F_d always opposes the motion:

$$m\ddot{x} + kx \pm F_d = F \sin \nu t.$$

If F_d is large compared to F, discontinuous motion will occur, but in most structures F_d is usually small so that an approximate continuous solution is valid. The approximate solution is obtained by linearising the equation of motion; this can be done by expressing F_d in terms of an equivalent viscous damping coefficient, c_d. From section 2.25,

$$c_d = \frac{4F_d}{\pi \nu X}$$

The solution to the linearised equation of motion gives the amplitude of the motion, X as;

$$X = \frac{F}{\sqrt{(k - m\nu^2)^2 + (c_d \nu)^2}}.$$

Thus $$X = \frac{F}{\sqrt{(k - m\nu^2)^2 + (4F_d/\pi X)^2}}.$$

That is, $$\frac{X}{X_s} = \frac{\sqrt{1 - (4F_d/\pi F)^2}}{1 - \left(\frac{\nu}{\omega}\right)^2}.$$

This expression is satisfactory for small damping forces, but breaks down if $4F_d/\pi F > 1$, that is if $F_d > (\pi/4)F$.

At resonance the amplitude is not limited by coulomb friction.

2.3.4 Response of a hysteretically damped structure to a simple harmonic exciting force with constant amplitude

In the single degree of freedom model shown in Fig. 2.18 the damping is hysteretic.

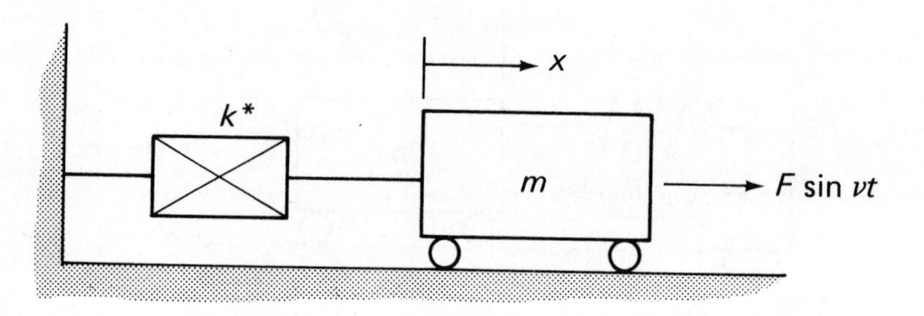

Fig. 2.18 – Single degree of freedom model of a forced structure with hysteretic damping.

The equation of motion is

$$m\ddot{x} + k^*x = F \sin \nu t.$$

Since $$k^* = k(1 + j\eta),$$

$$x = \frac{F \sin \nu t}{(k - m\nu^2) + j\eta k},$$

and $$\frac{X}{X_s} = \frac{1}{\sqrt{\left[1 - \left(\frac{\nu}{\omega}\right)^2\right]^2 + \eta^2}}$$

This result can also be obtained from the analysis of a viscous damped system by substituting $c = \eta k/\nu$.

It should be noted that if $c = \eta k/\nu$, at resonance $c = \eta\sqrt{km}$, that is $\eta = 2\zeta = 1/Q$.

2.3.5 Response of a structure to a suddenly applied force

Consider a single degree of freedom undamped system, such as the system shown in Fig. 2.19, which has been subjected to a suddenly applied force F.

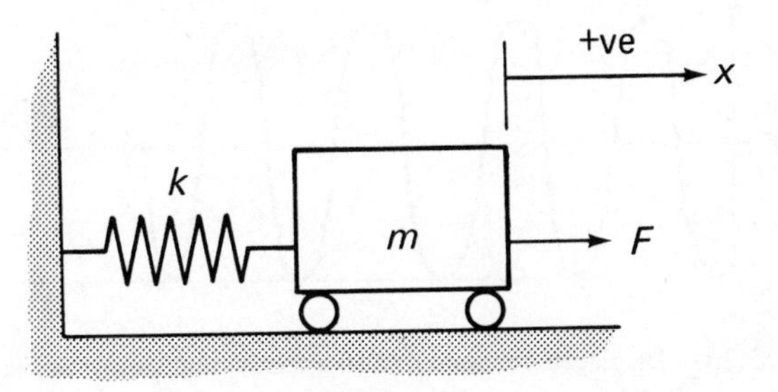

Fig. 2.19 – Single degree of freedom model with constant exciting force.

The equation of motion is $m\ddot{x} + kx = F$. The solution to this equation comprises a complementary function $A \sin \omega t + B \cos \omega t$, where $\omega = \sqrt{k/m}$ rad/s together with a particular solution. The particular solution may be found by using the D-operator. Thus the equation of motion can be written

$$\left(1 + \frac{D^2}{\omega^2}\right) x = \frac{F}{k},$$

and
$$x = \left(1 + \frac{D^2}{\omega^2}\right)^{-1} \frac{F}{k} = \frac{F}{k}.$$

That is, the complete solution to the equation of motion is

$$x = A \sin \omega t + B \cos \omega t + \frac{F}{k}.$$

If the initial conditions are such that $x = \dot{x} = 0$ at $t = 0$, then $B = -F/k$ and $A = 0$.

Thus $$x = \frac{F}{k}(1 - \cos \omega t).$$

The motion is shown in Fig. 2.20. It will be seen that the maximum dynamic displacement is twice the static displacement occurring under the same load. This is an important consideration in structures subjected to suddenly applied loads.

If the system possesses viscous damping of coefficient c, the solution to the equation of motion is $x = X\mathrm{e}^{-\zeta\omega t} \sin(\omega_v t + \phi) + F/k$.

With the same initial conditions as above,

$$x = \frac{F}{k}\left[1 - \frac{\mathrm{e}^{-\zeta\omega t}}{\sqrt{1-\zeta^2}} \sin\left(\omega\sqrt{1-\zeta^2}\,t + \tan^{-1}\frac{\sqrt{1-\zeta^2}}{\zeta}\right)\right].$$

This reduces to the undamped case if $\zeta = 0$. The response of the damped system is shown in Fig. 2.21.

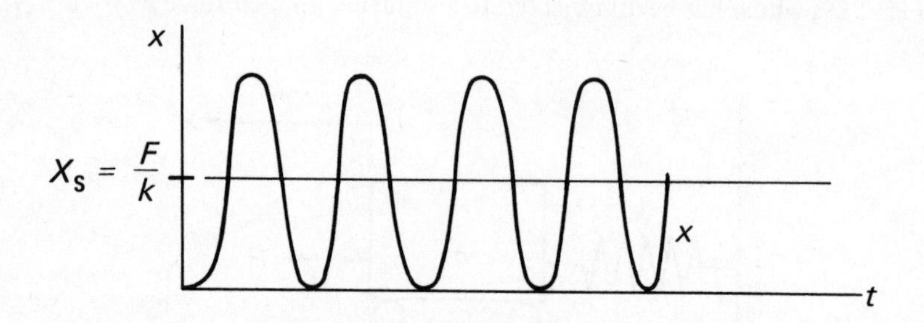

Fig. 2.20 – Displacement-time response for system shown in Fig. 2.19.

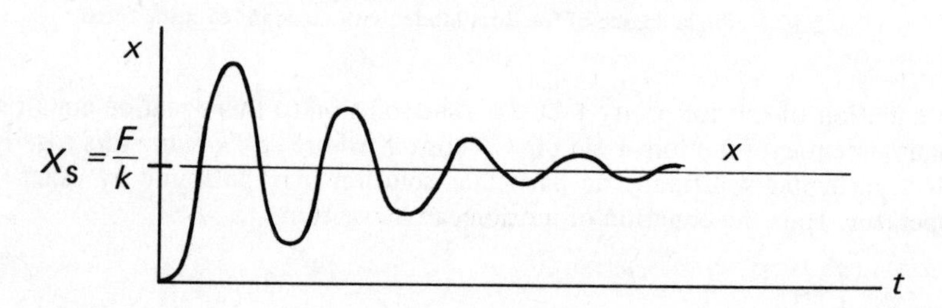

Fig. 2.21 – Displacement-time response for single degree of freedom system with viscous damping.

2.3.6 Shock excitation

Some structures are subjected to shock or impulse loads arising from suddenly applied, non-periodic, short duration exciting forces.

The impulsive force shown in Fig. 2.22 consists of a force of magnitude F_{max}/ϵ with a time duration of ϵ.

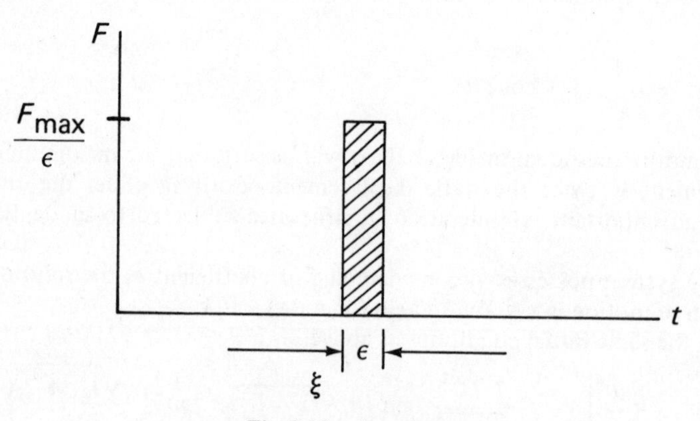

Fig. 2.22 – Impulse.

The impulse is equal to $\int_{t}^{t+\epsilon} \frac{F_{max}}{\epsilon} \, dt.$

When F_{max} is equal to unity, the force in the limiting case $\epsilon \to 0$ is called either the **unit impulse** or the **delta function**, and is identified by the symbol $\delta(t - \xi)$,

where $\int_0^{\infty} \delta(t - \xi) \, d\xi = 1.$

Since $F dt = m dv$, the impulse F_{max} acting on a body will result in a sudden change in its velocity without an appreciable change in its displacement. Thus the motion of a single degree of freedom system excited by an impulse F_{max} corresponds to free vibration with initial conditions $x(0) = 0$ and $\dot{x}(0) = v_0 = F_{max}/m$.

Once the response $g(t)$ say, to a unit impulse excitation is known, it is possible to establish the equation for the response of a system to an arbitrary exciting force $F(t)$. For this the arbitrary pulse is considered to comprise a series of impulses as shown in Fig. 2.23.

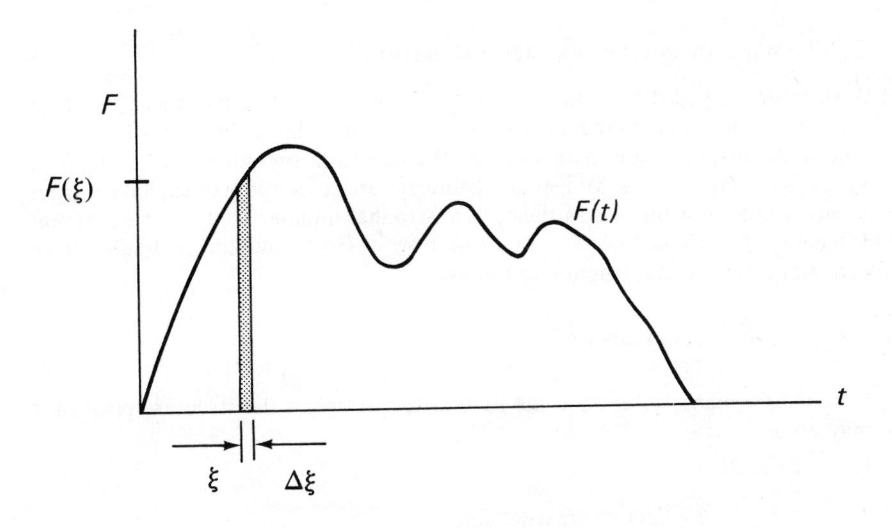

Fig. 2.23 – Force-time pulse.

If one of the impulses is examined which starts at time ξ, its magnitude is $F(\xi)\Delta\xi$, and its contribution to the system response at time t is found by replacing the time with the elapsed time $(t - \xi)$ as shown in Fig. 2.24.

If the system can be assumed to be linear, the principle of superposition can be applied, so that

$$x(t) = \int_0^t F(\xi) . \, g(t - \xi) . \, d\xi.$$

This is known as the **Duhamel integral.**

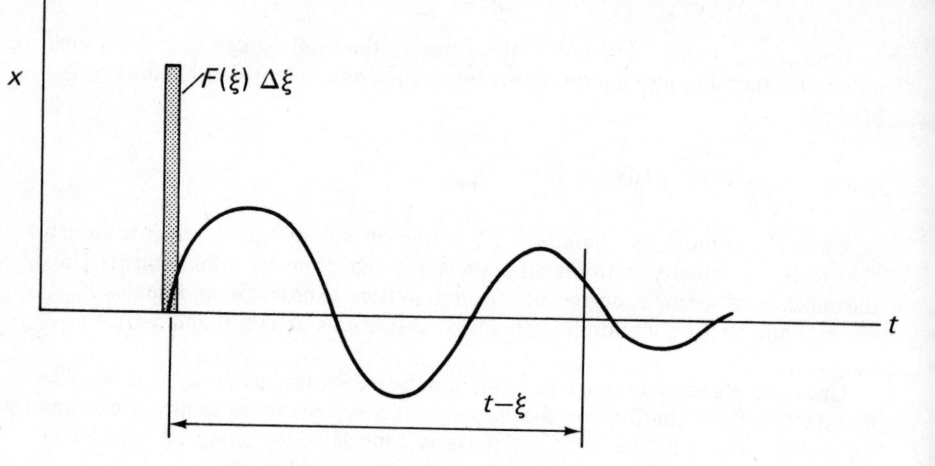

Fig. 2.24 – Displacement-time response to impulse.

2.3.7 Wind or current excited oscillation

A structure exposed to a fluid stream is subjected to an harmonically varying force in a direction perpendicular to the stream. This is because of eddy or vortex shedding on alternate sides of the structure on the leeward side. Tall structures such as masts, bridges and chimneys are susceptible to excitation from steady winds blowing across them. The **Strouhal number** relates the excitation frequency f_s, to the velocity of fluid flow v (m/s) and the hydraulic mean diameter D(m) of the structure as follows:

$$\text{Strouhal number} = \frac{f_s D}{v}\,.$$

If the frequency f_s is close the natural frequency of the structure resonance may occur.

For a structure,

$$D = \frac{4 \times \text{area of cross-section}}{\text{circumference}}\,,$$

so that for a chimney of circular cross-section and diameter d,

$$D = \frac{4\left(\dfrac{\pi}{4}\,d^2\right)}{\pi d} = d,$$

and for a building of rectangular cross-section $a \times b$,

$$D = \frac{4ab}{2(a+b)} = \frac{2ab}{(a+b)}\,.$$

Experimental evidence suggests a value of 0.2–0.24 for the Strouhal number for most flow rates and wind speeds encountered. This value is valid for Reynolds numbers in the range 3×10^5–3.5×10^6.

Example 6

For constructing a tanker terminal in a river estuary a number of cylindrical concrete piles were sunk into the river bed and left free standing. Each pile was 1 m diameter and protruded 20 m out of the river bed. The density of the concrete was 2400 kg/m^3 and the modulus of elasticity 14.10^6 kN/m^2. Estimate the velocity of the water flowing past a pile which will cause it to vibrate transversely to the direction of the current, assuming a pile to be a cantilever and taking a value for the Strouhal number

$$\frac{f_s D}{v} = 0.22,$$

where f_s is the frequency of flexural vibrations of a pile, D is the diameter and v is the velocity of the current.

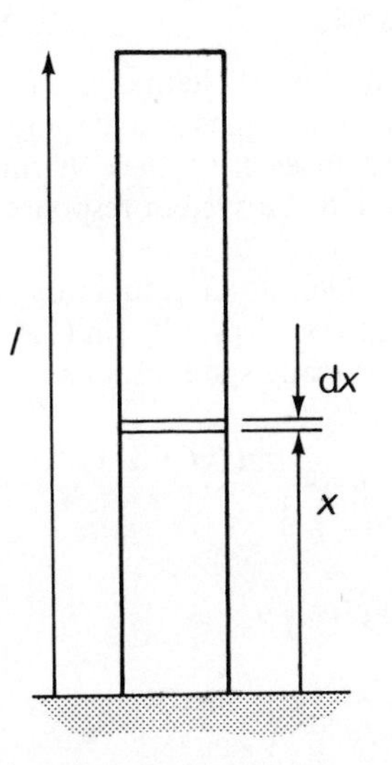

Consider the pile to be a cantilever of mass m, diameter D and length l, then the deflection y at a distance x from the root can be taken to be $y = y_l(1 - \cos \pi x/2l)$, where y_l is the deflection at the free end.

Thus $$\frac{d^2 y}{dx^2} = y_l . \left(\frac{\pi}{2l}\right)^2 . \cos \frac{\pi x}{2l}$$

$$\int_0^l EI \left(\frac{d^2 y}{dx^2}\right)^2 dx = EI \int_0^l y_l^2 \left(\frac{\pi}{2l}\right)^4 \cos^2 \frac{\pi x}{2l}\, dx$$

$$= EI.y_l^2\left(\frac{\pi}{2l}\right)^4.\frac{l}{2}$$

$$\int y^2\,dm = \int_0^l y_l^2\left(1-\cos\frac{\pi x}{2l}\right)^2\frac{m}{l}\,dx$$

$$= y_l^2.\frac{m}{l}\left(\frac{3}{2}-\frac{4}{\pi}\right)l$$

Hence
$$\omega^2 = \frac{EI.y_l^2.\left(\frac{\pi}{2l}\right)^4.\frac{l}{2}}{y_l^2\frac{m}{l}\left(\frac{3}{2}-\frac{4}{\pi}\right)l}.$$

Substituting numerical values gives $\omega = 5.53$ rad/s, i.e. $f = 0.88$ Hz. When $f_s = 0.88$ Hz resonance occurs, i.e. when $v = f_sD/0.22 = 0.88/0.22 = 4$ m/s.

2.3.8 Harmonic analysis

A function which is periodic but not harmonic can be represented by the sum of a number of terms, each term representing some multiple of the fundamental frequency. In a *linear* system each of these harmonic terms acts as if it alone were exciting the system, and the system response is the sum of the excitation of all the harmonics.

For example, if the periodic forcing function of a single degree of freedom undamped system is $F_1 \sin(vt + \alpha_1) + F_2 \sin(2vt + \alpha_2) + F_3 \sin(3vt + \alpha_3) + \ldots \ldots + F_n \sin(nvt + \alpha_n)$, the steady state response to $F_1 \sin(vt + \alpha_1)$ is

$$x_1 = \frac{F_1}{k\left(1-\left(\frac{v}{\omega}\right)^2\right)}\sin(vt+\alpha_1),$$

and the response to $F_2 \sin(2vt + \alpha_2)$ is

$$x_2 = \frac{F_2}{k\left(1-\left(\frac{2v}{\omega}\right)^2\right)}\sin(2vt+\alpha_2),$$

and so on, so that

$$x = \sum_{n=1}^{n}\frac{F_n}{k\left(1-\left(\frac{nv}{\omega}\right)^2\right)}\sin(nvt+\alpha_n).$$

Clearly that harmonic which is closest to the system natural frequency will most influence the response.

A periodic function can be written as the sum of a number of harmonic terms by writing a Fourier series for the function. A Fourier series can be written

$$F(t) = \frac{a_0}{2} + \sum_{n=1}^{\infty} (a_n \cos n\nu t + b_n \sin n\nu t),$$

where

$$a_0 = \frac{2}{\tau}\int_0^{\tau} F(t)\,\mathrm{d}t,$$

$$a_n = \frac{2}{\tau}\int_0^{\tau} F(t) \cos \nu t\,\mathrm{d}t, \text{ and}$$

$$b_n = \frac{2}{\tau}\int_0^{\tau} F(t) \sin \nu t\,\mathrm{d}t.$$

Thus, consider the first four terms of the Fourier series representation of the square wave shown in Fig. 2.25 to be required; $\tau = 2\pi$ so $\nu = 1$ rad/s.

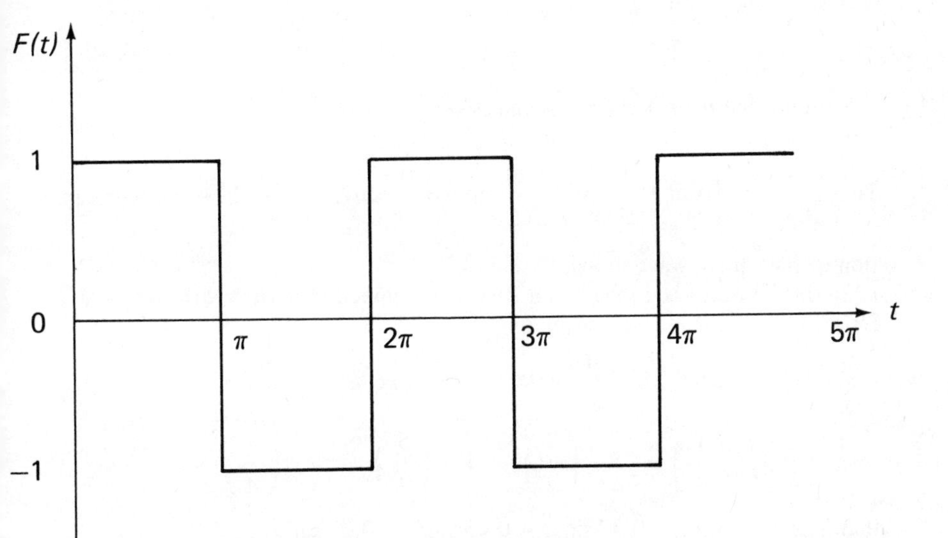

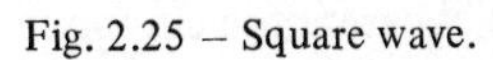

Fig. 2.25 – Square wave.

$$F(t) = \frac{a_0}{2} + a_1 \cos \nu t + a_2 \cos 2\nu t + \ldots$$

$$+ b_1 \sin \nu t + b_2 \sin 2\nu t + \ldots$$

$$a_0 = \frac{2}{\tau}\int_0^{\tau} F(t)\,\mathrm{d}t = \frac{2}{2\pi}\int_0^{\pi} 1\,\mathrm{d}t + \frac{2}{2\pi}\int_{\pi}^{2\pi} -1\,\mathrm{d}t = 0.$$

$$a_1 = \frac{2}{T}\int_0^T F(t)\cos\nu t\,dt$$

$$= \frac{2}{2\pi}\int_0^{\pi}\cos\nu t\,dt + \frac{2}{2\pi}\int_{\pi}^{2\pi} -\cos\nu t\,dt = 0.$$

Similarly $a_2 = a_3 = \ldots = 0.$

$$b_1 = \frac{2}{T}\int_0^T F(t)\sin\nu t\,dt$$

$$= \frac{2}{2\pi}\int_0^{\pi}\sin\nu t\,dt + \frac{2}{2\pi}\int_{\pi}^{2\pi} -\sin\nu t\,dt$$

$$= \frac{1}{\pi\nu}\left[-\cos\nu t\right]_0^{\pi} + \frac{1}{\pi\nu}\left[\cos\nu t\right]_{\pi}^{2\pi} = \frac{4}{\pi\nu}.$$

Since $\nu = 1$ rad/s,

$$b_1 = \frac{4}{\pi}.$$

It is found that $b_2 = 0, b_3 = \dfrac{4}{3\pi}$ and so on.

Thus $F(t) = \dfrac{4}{\pi}\left[\sin t + \dfrac{1}{3}\sin 3t + \dfrac{1}{5}\sin 5t + \dfrac{1}{7}\sin 7t \ldots\right]$ is the series representation of the square wave shown.

If this stimulus is applied to a simple undamped structure with $\omega = 4$ rad/s, the steady state response is given by

$$x = \frac{\frac{4}{\pi}\sin t}{1-\left(\frac{1}{4}\right)^2} + \frac{\frac{4}{3\pi}\sin 3t}{1-\left(\frac{3}{4}\right)^2} + \frac{\frac{4}{5\pi}\sin 5t}{1-\left(\frac{5}{4}\right)^2} + \frac{\frac{4}{7\pi}\sin 7t}{1-\left(\frac{7}{4}\right)^2}\ldots$$

that is, $x = 1.36\sin t + 0.97\sin 3t - 0.45\sin 5t - 0.09\sin 7t\ldots.$

Usually three or four terms of the series dominate the predicted response.

It is worth sketching the components of $F(t)$ above to show that they produce a reasonable square wave, whereas the components of x do not.

Chapter 3

The vibration of structures with more than one degree of freedom

Most real structures have several bodies and several restraints and therefore several degrees of freedom. The number of **degrees of freedom** that a structure possesses is equal to the number of independent coordinates necessary to describe the motion of the structure. Since no body is completely rigid, and no spring is without mass, every real structure has many degrees of freedom, and sometimes it is not sufficiently realistic to approximate a structure by a single degree of freedom model. Therefore it is necessary to study the vibration of structures with more than one degree of freedom.

Each flexibly connected body in a multi-degree of freedom structure can move independently of the other bodies, and only under certain conditions will all bodies undergo an harmonic motion at the same frequency. Since all bodies move with the same frequency, they all attain their maximum displacement at the same time, even if they do not all move in the same direction. When such motion occurs the frequency is called a **natural frequency** of the structure, and the motion is a **principal mode** of vibration: the number of natural frequencies and principal modes that a structure possesses is equal to the number of degrees of freedom of that structure. The deployment of the structure at its lowest or first natural frequency is called its first **mode**, at the next highest or second natural frequency it is called the second mode and so on.

A two degree of freedom model of a structure will be considered first. This is because the addition of more degrees of freedom increases the labour of the solution procedure but does not introduce any new analytical principles.

Initially, we will obtain the equations of motion for a two degree of freedom model, and from these find the natural frequencies and corresponding mode shapes.

Examples of structures requiring a two degree of freedom model are a two-storey shear frame building as shown in Fig. 3.1, and a vehicle where translation motion coupled with rotation in the same plane takes place, Fig. 3.2.

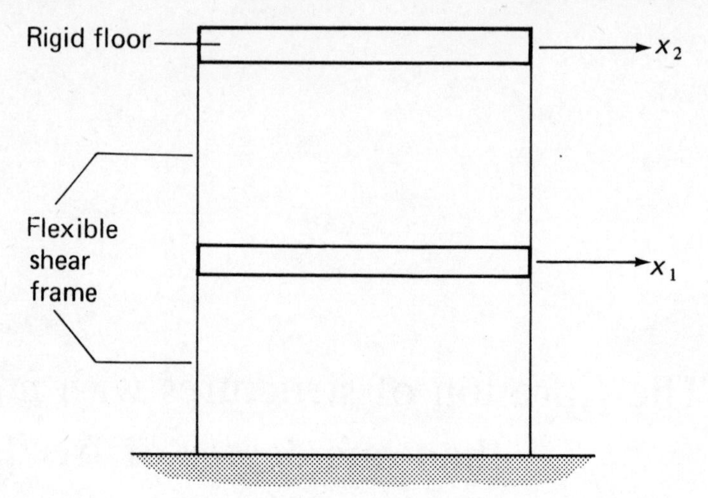

Fig. 3.1 – Two degree of freedom shear frame building model.

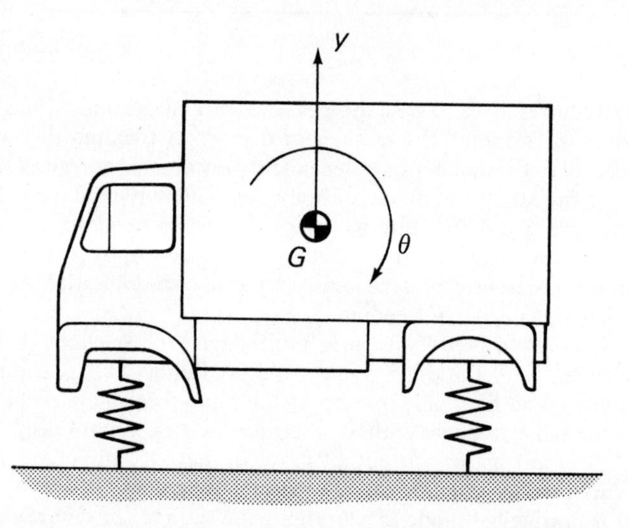

Fig. 3.2 – Two degree of freedom model, rotation plus translation.

3.1 STRUCTURES WITH TWO DEGREES OF FREEDOM

3.1.1 Free vibration of an undamped structure

If a two degree of freedom model is appropriate for a structure, the model may take a variety of forms, for example those shown in Figs. 3.3(a) and (b). The equations of motion for both of these models are

$$m_1\ddot{x}_1 = -k_1x_1 - k(x_1 - x_2),$$

and $$m_2\ddot{x}_2 = k(x_1 - x_2) - k_2x_2.$$

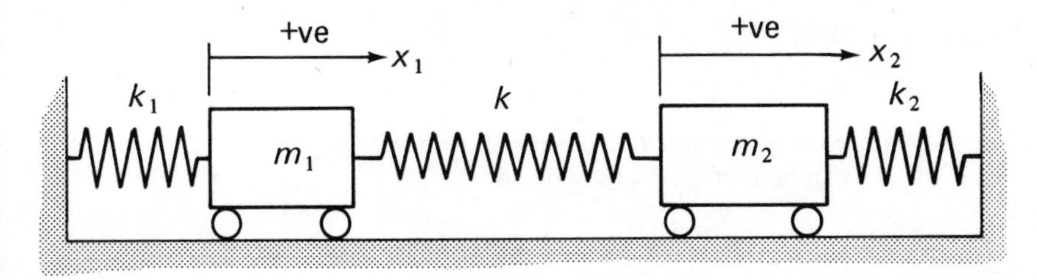

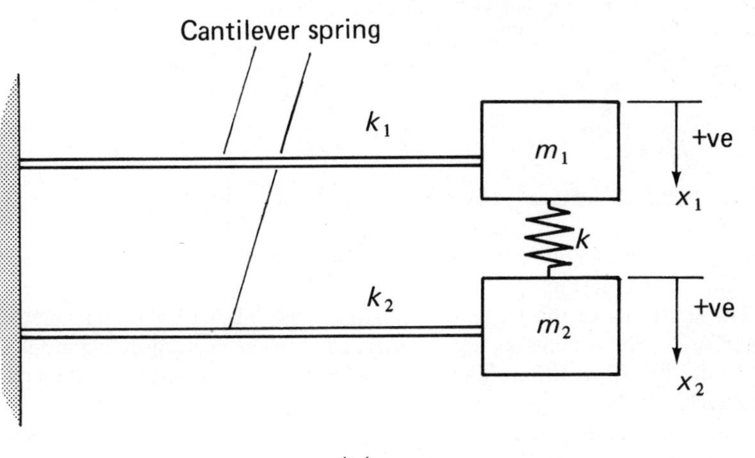

(b)

Fig. 3.3 – Two degree of freedom model, translation vibration.

These equations can be solved for the natural frequencies and corresponding mode shapes, by assuming solutions of the form

$$x_1 = A_1 \sin(\omega t + \psi),$$

and $$x_2 = A_2 \sin(\omega t + \psi).$$

This assumes that x_1 and x_2 oscillate with the same frequency ω and are either in phase or π out of phase. This is a sufficient condition to make ω a natural frequency.

Substituting these solutions into the equations of motion gives,

$$-m_1 A_1 \omega^2 \sin(\omega t + \psi) = -k_1 A_1 \sin(\omega t + \psi) - k(A_1 - A_2) \sin(\omega t + \psi),$$

and $$-m_2 A_2 \omega^2 \sin(\omega t + \psi) = k(A_1 - A_2) \sin(\omega t + \psi) - k_2 A_2 \sin(\omega t + \psi).$$

That is, $A_1(k+k_1-m_1\omega^2)+A_2(-k)=0$,

and $A_1(-k)+A_2(k_2+k-m_2\omega^2)=0$,

since the equations of motion are true for all values of t.

A_1 and A_2 can be eliminated by writing

$$\begin{vmatrix} k+k_1-m_1\omega^2 & -k \\ -k & k+k_2-m_2\omega^2 \end{vmatrix} = 0.$$

This is the **characteristic** or **frequency equation**. Alternatively, we may write

$$\frac{A_1}{A_2} = \frac{k}{k+k_1-m_1\omega^2}$$

and

$$\frac{A_1}{A_2} = \frac{k_2+k-m_2\omega^2}{k}.$$

Thus

$$\frac{k}{k+k_1-m_1\omega^2} = \frac{k_2+k-m_2\omega^2}{k},$$

and

$$(k+k_1-m_1\omega^2)(k_2+k-m_2\omega^2)-k^2=0;$$

this result is the frequency equation and could also be obtained by multiplying out the above determinant.

The solutions to the frequency equation give the natural frequencies of free vibration of the systems in Figs. 3.3(a) and (b). The corresponding mode shapes are found by substituting these frequencies, in turn, into either of the equations for A_1/A_2.

For example, consider the case when $k_1=k_2=k$, and $m_1=m_2=m$. The frequency equation is $(2k-m\omega^2)^2-k^2=0$, that is, $m^2\omega^4-4mk\omega^2+3k^2=0$.

Thus $\omega_1=\sqrt{\frac{k}{m}}$ rad/s, and $\omega_2=\sqrt{\frac{3k}{m}}$ rad/s.

If $\omega=\sqrt{\frac{k}{m}}$ rad/s, $\left(\frac{A_1}{A_2}\right)_{\omega=\sqrt{k/m}}=+1$,

and if $\omega=\sqrt{\frac{3k}{m}}$ rad/s, $\left(\frac{A_1}{A_2}\right)_{\omega=\sqrt{3k/m}}=-1$.

This gives the mode shapes corresponding to the frequencies ω_1 and ω_2.

Thus the first mode of free vibration occurs at a frequency $1/2\pi\sqrt{k/m}$ Hz, and then $(A_1/A_2)^{\mathrm{I}}=1$; that is the bodies move in phase with each other and with the same amplitude, as if connected by a rigid link, as shown in Fig. 3.4. The second mode of free vibration occurs at a frequency $1/2\pi\sqrt{3k/m}$ Hz and then $(A_1/A_2)^{\mathrm{II}}=-1$, so that the bodies move exactly out of phase with each other, but with the same amplitude. This mode is also shown in Fig. 3.4.

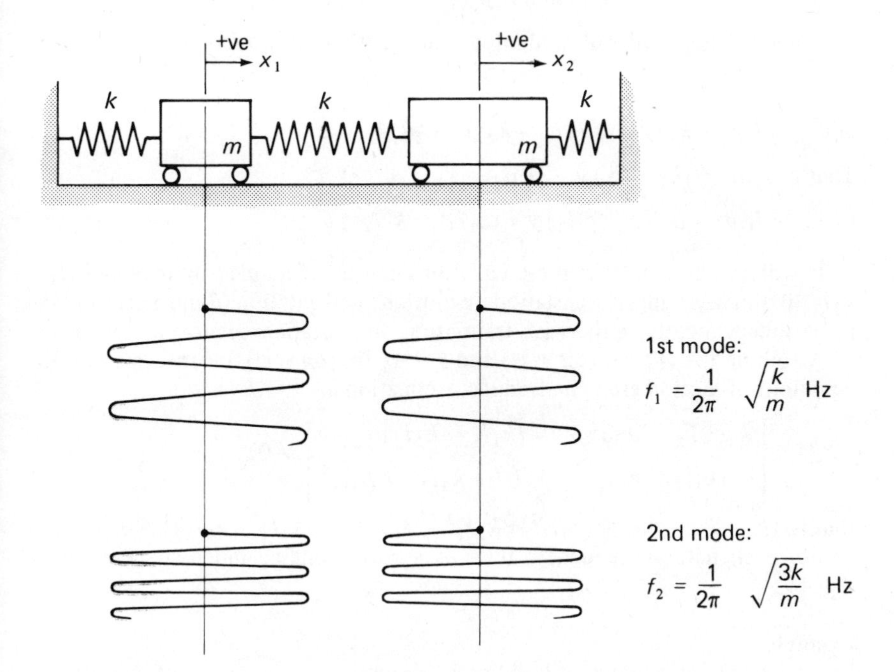

Fig. 3.4 – Natural frequencies and mode shapes for two degree of freedom translation vibration system.

3.1.2 Coordinate coupling

In some structures, such as that shown in Fig. 3.2, the coordinates of motion are coupled in the equations of motion, because in general, translation and rotation occur simultaneously. Consider the model shown in Fig. 3.5, only motion in the plane of the figure is considered, horizontal motion being neglected because the lateral stiffness of the springs is assumed negligible. The coordinates of rotation, θ and translation, y are coupled as shown in the figure. G is the centre of mass of the body of mass m and moment of inertia about G, I_G.

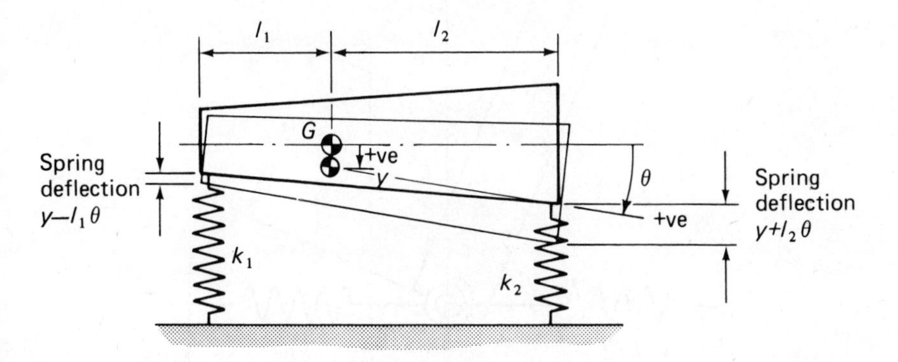

Fig. 3.5 – Two degree of freedom model, rotation plus translation.

For small amplitudes of oscillation, the equations of motion are;

$$m\ddot{y} = -k_1(y - l_1\theta) - k_2(y + l_2\theta),$$

and $$I_G\ddot{\theta} = k_1(y - l_1\theta)l_1 - k_2(y + l_2\theta)l_2.$$

That is, $$m\ddot{y} + (k_1 + k_2)y - (k_1l_1 - k_2l_2)\theta = 0,$$

and $$I_G\ddot{\theta} - (k_1l_1 - k_2l_2)y + (k_1l_1^2 + k_2l_2^2)\theta = 0.$$

It will be observed that these equations can be uncoupled by making $k_1l_1 = k_2l_2$, if this is arranged translation (y-motion) and rotation (θ-motion) can take place independently. Otherwise translation and rotation occur simultaneously.

Assuming $y = A_1 \sin(\omega t + \psi)$ and $\theta = A_2 \sin(\omega t + \psi)$, substituting into the equations of motion gives the frequency equation as,

$$\begin{vmatrix} k_1 + k_2 - m\omega^2 & -(k_1l_1 - k_2l_2) \\ -(k_1l_1 - k_2l_2) & k_1l_1^2 + k_2l_2^2 - I_G\omega^2 \end{vmatrix} = 0,$$

that is, $(k_1 + k_2 - m\omega^2)(k_1l_1^2 + k_2l_2^2 - I_G\omega^2) - (k_1l_1 - k_2l_2)^2 = 0.$

For each natural frequency, there is a corresponding mode shape given by A_1/A_2.

Example 7

In a study of earthquakes, a building is idealised as a rigid body of mass M supported on two springs, one giving translational stiffness k and the other rotational stiffness k_T, as shown.

If I_G is the mass moment of inertia of the building about its mass centre G, write down the equations of motion using coordinates x for the translation from the equilibrium position, and θ for the rotation of the building.

Hence determine the frequency equation of the motion.

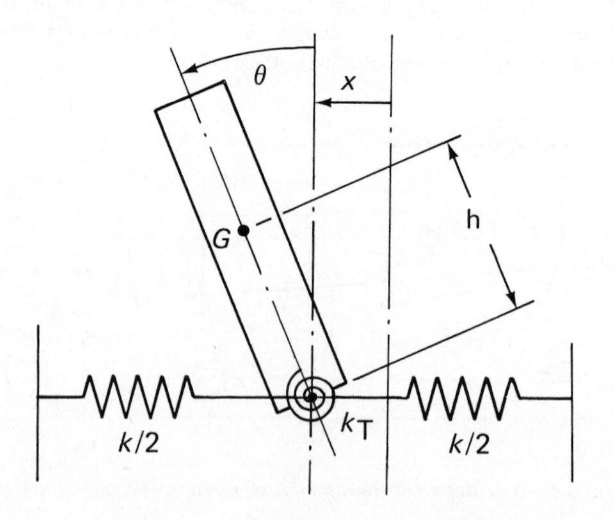

The FBD's are as follows.

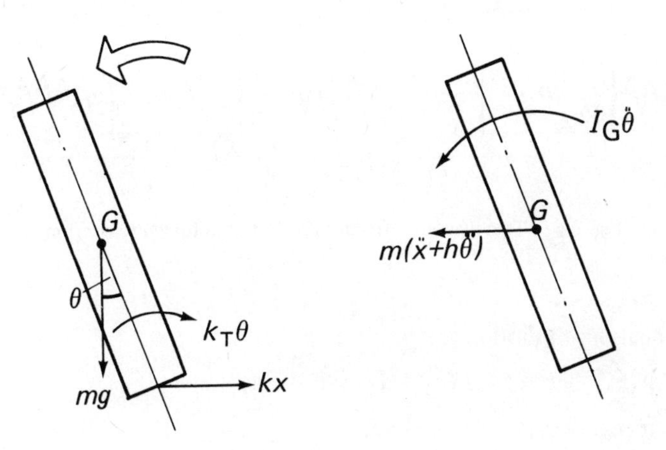

Assume small θ (earthquakes), hence

$$m(\ddot{x} + h\ddot{\theta}) = -kx,$$

and $$I_G\ddot{\theta} + m(\ddot{x} + h\ddot{\theta})\,h = -k_T\theta + mgh\theta.$$

The equations of motion are therefore,

$$mh\ddot{\theta} + m\ddot{x} + kx = 0,$$

and $$mh\ddot{x} + (mh^2 + I_G)\,\ddot{\theta} - (mgh - k_T)\,\theta = 0.$$

If $$\theta = A_1 \sin \omega t \quad \text{and} \quad x = A_2 \sin \omega t,$$

$$-\,mh\omega^2 A_1 - m\omega^2 A_2 + kA_2 = 0,$$

and $$+\,mh\omega^2 A_2 + (mh^2 + I_G)\,\omega^2 A_1 + (mgh - k_T)\,A_1 = 0.$$

The frequency equation is:

$$\begin{vmatrix} -mh\omega^2 & k - m\omega^2 \\ (mh^2 + I_G)\,\omega^2 + (mgh - k_T) & mh\omega^2 \end{vmatrix} = 0,$$

That is $$(mh\omega^2)^2 + (k - m\omega^2)\,[(mh^2 + I_G)\,\omega^2 + (mgh - k_T)] = 0,$$

or, $$mI_G\omega^4 - \omega^2\,[mkh^2 + I_Gk - m^2gh + mk_T] - mghk + kk_T = 0.$$

3.1.3 Forced vibration

Harmonic excitation of vibration in a structure may be generated in a number of ways, for example by unbalanced rotating or reciprocating machinery, or it may arise from periodic excitation containing a troublesome harmonic component.

A two degree of freedom model of a structure excited by an harmonic force $F \sin \nu t$ is shown in Fig. 3.6. Damping is assumed to be negligible. The force has a constant amplitude F and a frequency $\nu/2\pi$ Hz.

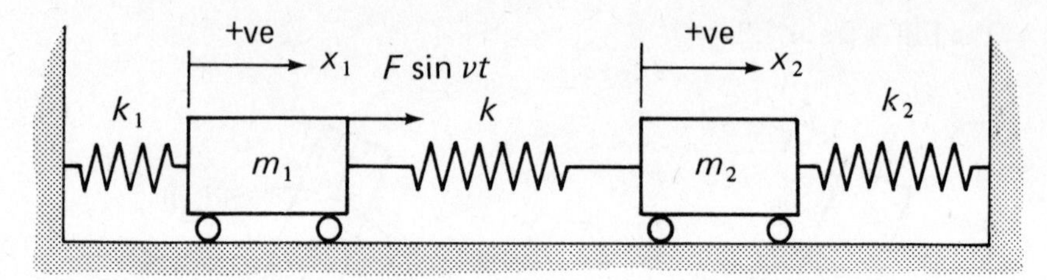

Fig. 3.6 – Two degree of freedom model with forced excitation.

The equations of motion are:

$$m_1\ddot{x}_1 = -k_1x_1 - k(x_1 - x_2) + F\sin\nu t,$$

and $$m_2\ddot{x}_2 = k(x_1 - x_2) - k_2x_2.$$

Since there is zero damping, the motions are either in phase or π out of phase with the driving force, so that the following solutions may be assumed:

$$x_1 = A_1\sin\nu t \quad\text{and}\quad x_2 = A_2\sin\nu t.$$

Substituting these solutions into the equations of motion gives:

$$A_1(k_1 + k - m_1\nu^2) + A_2(-k) = F,$$

and $$A_1(-k) + A_2(k_2 + k - m_2\nu^2) = 0.$$

Thus $$A_1 = \frac{F(k_2 + k - m_2\nu^2)}{\Delta},$$

and $$A_2 = \frac{Fk}{\Delta},$$

where $$\Delta = (k_2 + k - m_2\nu^2)(k_1 + k - m_1\nu^2) - k^2,$$

and $\Delta = 0$ is the frequency equation.

Hence the response of the structure to the exciting force is determined.

Example 8

A two-wheel trailer is drawn over an undulating surface in such a way that the vertical motion of the tyre may be regarded as sinusoidal, the pitch of the undulations being 5 m. The combined stiffness of the tyres is 170 kN/m and that of the main springs is 60 kN/m; the axle and attached parts have a mass of 400 kg and the mass of the body is 500 kg. Find (a) the critical speeds of the trailer in km/h, and (b) the amplitude of the trailer body vibration if the trailer is drawn at 50 km/h, and the amplitude of the undulations is 0.1 m.

The equations of motion are:

$$m_1\ddot{x}_1 = -k_1(x_1 - x_2),$$

and $$m_2\ddot{x}_2 = k_1(x_1 - x_2) - k_2(x_2 - x_3).$$

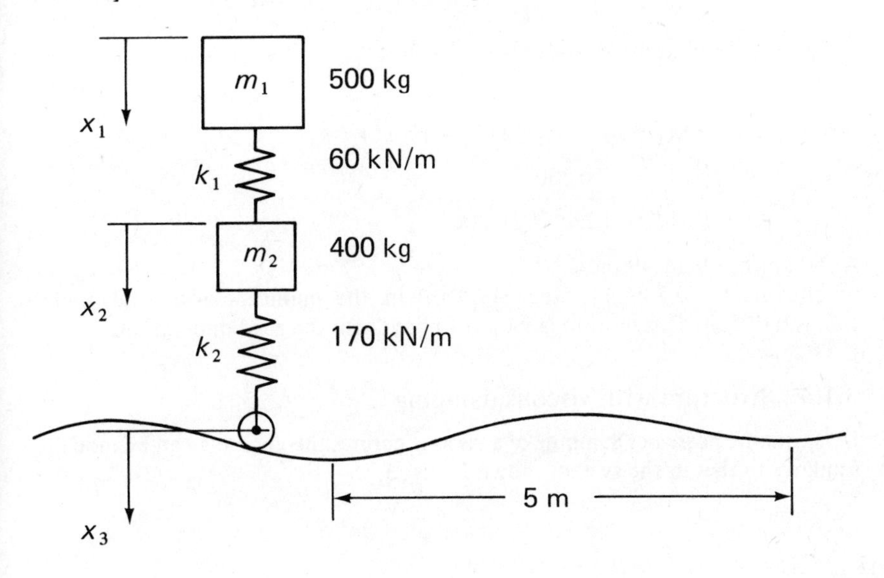

Assuming $x_1 = A_1 \sin \nu t$, $x_2 = A_2 \sin \nu t$, and $x_3 = A_3 \sin \nu t$

$$A_1(k_1 - m_1\nu^2) + A_2(-k_1) = 0,$$

and
$$A_1(-k_1) + A_2(k_1 + k_2 - m_2\nu^2) = k_2 A_3.$$

The frequency equation is:

$$(k_1 + k_2 - m_2\nu^2)(k_1 - m_1\nu^2) - k_1^2 = 0.$$

The critical speeds are those which correspond to the natural frequencies and hence excite resonances.

The frequency equation simplifies to

$$m_1 m_2 \nu^4 - (m_1 k_1 + m_1 k_2 + m_2 k_1)\nu^2 + k_1 k_2 = 0.$$

Hence, substituting the given data,

$$500 \times 400 \times \nu^4 - (500 \times 60 + 500 \times 170 + 400 \times 60)\, 10^3 \nu^2 + 60 \times 170 \times 10^6 = 0.$$

That is $0.2\nu^4 - 139\nu^2 + 10\,200 = 0$.

Thus $\nu = 16.3$ rad/s or 20.78 rad/s, and $f = 2.59$ Hz or 3.3 Hz.

Now if the trailer is drawn at v km/h, or $v/3.6$ m/s, the frequency is $v/(3.6 \times 5)$ Hz.

Therefore the critical speeds are:

$$v_1 = 18 \times 2.59 = 46.6 \text{ km/h},$$

and
$$v_2 = 18 \times 3.3 = 59.4 \text{ km/h}.$$

Towing the trailer at either of these speeds will excite resonance in the system.

From the equations of motion,

$$A_1 = \left\{ \frac{k_1 k_2}{(k_1 + k_2 - m_2 \nu^2)(k_1 - m_1 \nu^2) - k_1^2} \right\} A_3,$$

$$= \left\{ \frac{10\,200}{0.2\nu^4 - 139\nu^2 + 10\,200} \right\} A_3.$$

At 50 km/h, $\nu = 17.49$ rad/s.

Thus $A_1 = -0.749\, A_3$. Since $A_3 = 0.1$ m, the amplitude of the trailer vibration is 0.075 m. This motion is π out of phase with the road undulations.

3.1.4 Structure with viscous damping

If a structure possesses damping of a viscous nature, the damping can be modelled similarly to that in the system shown in Fig. 3.7.

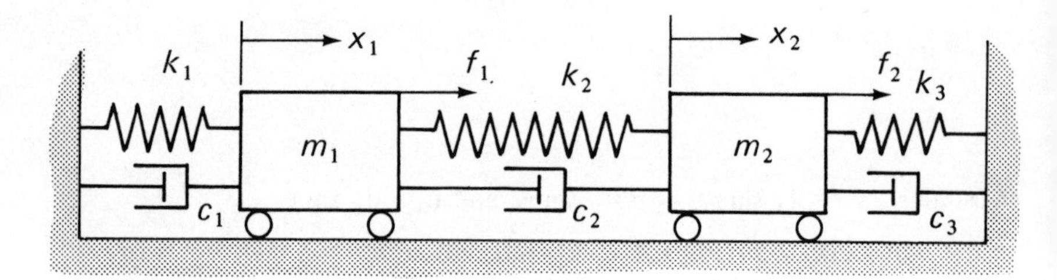

Fig. 3.7 – Two degree of freedom viscous damped model with forced excitation.

For this system the equations of motion are:

$$m_1 \ddot{x}_1 + k_1 x_1 + k_2 (x_1 - x_2) + c_1 \dot{x}_1 + c_2 (\dot{x}_1 - \dot{x}_2) = f_1,$$

and $$m_2 \ddot{x}_2 + k_2 (x_2 - x_1) + k_3 x_2 + c_2 (\dot{x}_2 - \dot{x}_1) + c_3 \dot{x}_2 = f_2.$$

Solutions of the form $x_1 = A_1 e^{st}$ and $x_2 = A_2 e^{st}$ can be assumed, where the Laplace operator s is equal to $a + jb$, $j = \sqrt{-1}$, and a and b are real—that is each solution contains an harmonic component of frequency b, and a vibration decay component of damping factor a. By substituting these solutions into the equations of motion a frequency equation of the form

$$s^4 + \alpha s^3 + \beta s^2 + \gamma s + \delta = 0$$

can be deduced, where α, β, γ and δ are real coefficients. From this equation four roots and thus four values of s can be obtained. In general the roots form two complex conjugate pairs such as $a_1 \pm jb_1$, and $a_2 \pm jb_2$. These represent solutions of the form $x = \text{Real } X e^{at} . e^{jbt} = X e^{at} \cos bt$. That is, the motion of the bodies is harmonic, and decays exponentially with time. The parameters of the structure determine the magnitude of the frequency and the decay rate.

3.2 STRUCTURES WITH MORE THAN TWO DEGREES OF FREEDOM

The vibration analysis of a structure with three or more degrees of freedom can be carried out in the same way as the analysis given above for two degrees of freedom. However, the method becomes tedious for many degrees of freedom, and numerical methods may have to be used to solve the frequency equation. A computer can, of course, be used to solve the frequency equation and determine the corresponding mode shapes. Although computational and computer techniques are extensively used in the analysis of multi-degree of freedom structures, it is essential for the analytical and numerical bases of any program used to be understood, to ensure its relevance to the problem considered, and that the program does not introduce unacceptable approximations and calculation errors. For this reason it is necessary to derive the basic theory and equations for multi-degree of freedom structures. Computational techniques are essential, and widely used, for the analysis of the sophisticated structural models often devised and considered necessary, and computer packages are available for routine analyses. However, considerable economies in writing the analysis and performing the computations can be achieved, by adopting a matrix method for the analysis. Alternatively an energy solution can be obtained by using the Lagrange Equation, or some simplification in the analysis achieved by using the receptance technique. The matrix method will be considered first.

3.2.1 The matrix method

The matrix method for analysis is a convenient way of handling several equations of motion. Furthermore, specific information about a structure, such as its lowest natural frequency, can be obtained without carrying out a complete and detailed analysis. The matrix method of analysis is particularly important because it forms the basis of many computer solutions to vibration problems. The method can best be demonstrated by means of an example. For a full description of the matrix method see *Mechanical Vibrations: An Introduction to Matrix Methods*, by J. M. Prentis and F. A. Leckie (Longmans).

Example 9

A structure is modelled by the three degree of freedom system shown. Determine the highest natural frequency of free vibration and the associated mode shape.

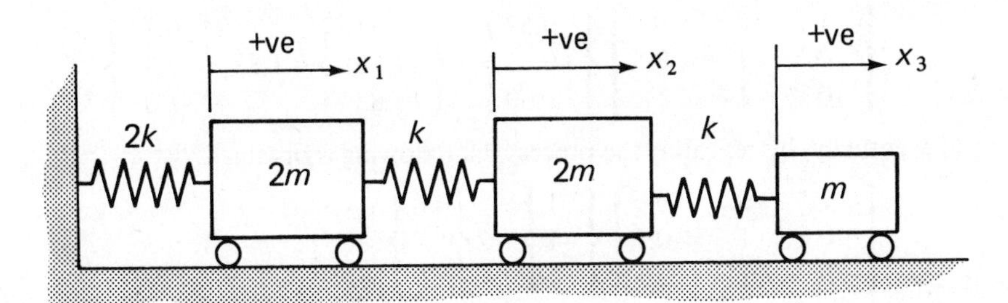

The equations of motion are

$$2m\ddot{x}_1 + 2kx_1 + k(x_1 - x_2) = 0,$$

$$2m\ddot{x}_2 + k(x_2 - x_1) + k(x_2 - x_2) = 0,$$

and $$m\ddot{x}_3 + k(x_3 - x_2) = 0.$$

If x_1, x_2 and x_3 take the form $X \sin \omega t$ and $\lambda = m\omega^2/k$, these equations can be written:

$$\tfrac{3}{2}X_1 - \tfrac{1}{2}X_2 \qquad\qquad = \lambda X_1,$$

$$\tfrac{1}{2}X_1 + \quad X_2 - \tfrac{1}{2}X_3 = \lambda X_2,$$

and $$- \quad X_2 + \quad X_3 = \lambda X_3.$$

that is, $$\begin{bmatrix} 1.5 & -0.5 & 0 \\ -0.5 & 1 & -0.5 \\ 0 & -1 & 1 \end{bmatrix} \begin{Bmatrix} X_1 \\ X_2 \\ X_3 \end{Bmatrix} = \lambda \begin{Bmatrix} X_1 \\ X_2 \\ X_3 \end{Bmatrix}$$

or $$[\mathbf{S}]\,\{\mathbf{X}\} = \lambda\{\mathbf{X}\}$$

where $[\mathbf{S}]$ is the **system matrix**, $\{\mathbf{X}\}$ is a column matrix and the factor λ is a scalar quantity.

This matrix equation can be solved by an iteration procedure. This procedure is started by assuming a set of deflections for the column matrix $\{\mathbf{X}\}$ and multiplying by $[\mathbf{S}]$; this results in a new column matrix. This matrix is normalised by making one of the amplitudes unity and dividing each term in the column by the particular amplitude which was put equal to unity. The procedure is repeated until the amplitudes stabilise to a definite pattern. Convergence is always to the highest value of λ and its associated column matrix. Since $\lambda = m\omega^2/k$, this means that the highest natural frequency is found. Thus to start the iteration a reasonable assumed mode would be:

$$\begin{Bmatrix} X_1 \\ X_2 \\ X_3 \end{Bmatrix} = \begin{Bmatrix} 1 \\ -1 \\ 2 \end{Bmatrix}.$$

Now, $$\begin{bmatrix} 1.5 & -0.5 & 0 \\ -0.5 & 1 & -0.5 \\ 0 & -1 & 1 \end{bmatrix} \begin{Bmatrix} 1 \\ -1 \\ 2 \end{Bmatrix} = \begin{Bmatrix} 2 \\ -2.5 \\ 3 \end{Bmatrix} = 3 \begin{Bmatrix} 0.67 \\ -0.83 \\ 1 \end{Bmatrix}$$

Using this new column matrix gives:

$$\begin{bmatrix} 1.5 & -0.5 & 0 \\ -0.5 & 1 & -0.5 \\ 0 & -1 & 1 \end{bmatrix} \begin{Bmatrix} 0.67 \\ -0.83 \\ 1.00 \end{Bmatrix} = \begin{Bmatrix} 1.415 \\ -1.665 \\ 1.83 \end{Bmatrix} = 1.83 \begin{Bmatrix} 0.77 \\ -0.91 \\ 1.0 \end{Bmatrix}$$

and eventually, by repeating the process the following is obtained:

$$\begin{bmatrix} 1.5 & -0.5 & 0 \\ -0.5 & 1 & -0.5 \\ 0 & -1 & 1 \end{bmatrix} \begin{Bmatrix} 1 \\ -1 \\ 1 \end{Bmatrix} = 2 \begin{Bmatrix} 1 \\ -1 \\ 1 \end{Bmatrix}$$

Hence $\lambda = 2$ and $\omega^2 = 2k/m$. λ is an **eigenvalue** of **[S]**, and the associated value of $\{\mathbf{X}\}$ is the corresponding **eigenvector** of **[S]**. The eigenvector gives the mode shape.

Thus the highest natural frequency is $1/2\pi \sqrt{2k/m}$ Hz, and the associated mode shape is 1:−1:1. Thus if $X_1 = 1, X_2 = -1$ and $X_3 = 1$.

If the lowest natural frequency is required, it can be found from the lowest eigenvalue. This can be obtained directly by inverting **[S]** and premultiplying $[\mathbf{S}]\{\mathbf{X}\} = \lambda\{\mathbf{X}\}$ by $\lambda^{-1}[\mathbf{S}]^{-1}$.

Thus $[\mathbf{S}]^{-1}\{\mathbf{X}\} = \lambda^{-1}\{\mathbf{X}\}$. Iteration of this equation yields the largest value of λ^{-1} and hence the lowest natural frequency. A reasonable assumed mode for the first iteration would be

$$\begin{Bmatrix} 1 \\ 1 \\ 2 \end{Bmatrix}.$$

Alternatively, the lowest eigenvalue can be found from the **flexibility matrix.** The flexibility matrix is written in terms of the influence coefficients. The **influence coefficient** α_{pq} of a system is the deflection (or rotation) at the point p due to a unit force (or moment) applied at a point q. Thus, since the force each body applies is the product of its mass and acceleration,

$$X_1 = \alpha_{11}\, 2mX_1\omega^2 + \alpha_{12}\, 2mX_2\omega^2 + \alpha_{13}\, mX_3\omega^2,$$

$$X_2 = \alpha_{21}\, 2mX_1\omega^2 + \alpha_{22}\, 2mX_2\omega^2 + \alpha_{23}\, mX_3\omega^2,$$

and

$$X_3 = \alpha_{31}\, 2mX_1\omega^2 + \alpha_{32}\, 2mX_2\omega^2 + \alpha_{33}\, mX_3\omega^2.$$

or

$$\begin{bmatrix} 2\alpha_{11} & 2\alpha_{12} & \alpha_{13} \\ 2\alpha_{21} & 2\alpha_{22} & \alpha_{23} \\ 2\alpha_{31} & 2\alpha_{32} & \alpha_{33} \end{bmatrix} \begin{Bmatrix} X_1 \\ X_2 \\ X_3 \end{Bmatrix} = \frac{1}{m\omega^2} \begin{Bmatrix} X_1 \\ X_2 \\ X_3 \end{Bmatrix}.$$

The influence coefficients are calculated by applying a unit force or moment to each body in turn. Since the same unit force acts between the support and its point of application, the displacement of the point of application of the force is the sum of the extensions of the springs extended. The displacements of all points beyond the point of application of the force are the same.

Thus $\alpha_{11} = \alpha_{12} = \alpha_{13} = \alpha_{21} = \alpha_{31} = \dfrac{1}{2k}$,

$$\alpha_{22} = \alpha_{23} = \alpha_{32} = \frac{1}{2k} + \frac{1}{k} = \frac{3}{2k},$$

and

$$\alpha_{33} = \frac{1}{2k} + \frac{1}{k} + \frac{1}{k} = \frac{5}{2k}.$$

Iteration causes the eigenvalue $k/m\omega^2$ to converge to its highest value, and hence the lowest natural frequency is found. The other natural frequencies of the system can be found by applying the orthogonality relation between the principal modes of vibration.

3.2.1.1 Orthogonality of the principal modes of vibration

Consider a linear elastic system that has n degrees of freedom, n natural frequencies and n principal modes.

The orthogonality relation between the principal modes of vibration for an n degree of freedom system is:

$$\sum_{i=1}^{n} m_i A_i(r)\, A_i(s) = 0,$$

where $A_i(r)$ are the amplitudes corresponding to the rth mode,

and $A_i(s)$ are the amplitudes corresponding to the sth mode.

This relationship is used to sweep unwanted modes from the system matrix, as illustrated in the following example.

Example 10

Consider the three degree of freedom model of a structure shown.

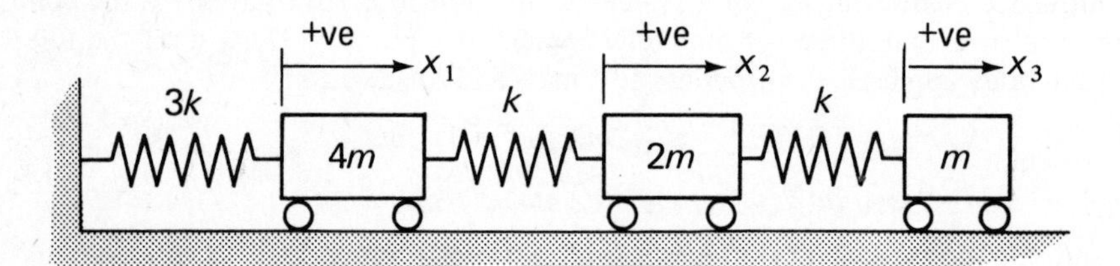

The equations of motion in terms of the influence coefficients are:

$$X_1 = 4\alpha_{11}\, mX_1\omega^2 + 2\alpha_{12}\, mX_2\omega^2 + \alpha_{13}\, mX_3\omega^2,$$

$$X_2 = 4\alpha_{21}\, mX_1\omega^2 + 2\alpha_{22}\, mX_2\omega^2 + \alpha_{23}\, mX_3\omega^2,$$

and $$X_3 = 4\alpha_{31}\, mX_1\omega^2 + 2\alpha_{32}\, mX_2\omega^2 + \alpha_{33}\, mX_3\omega^2.$$

That is, $$\begin{Bmatrix} X_1 \\ X_2 \\ X_3 \end{Bmatrix} = \omega^2 m \begin{bmatrix} 4\alpha_{11} & 2\alpha_{12} & \alpha_{13} \\ 4\alpha_{21} & 2\alpha_{22} & \alpha_{23} \\ 4\alpha_{31} & 2\alpha_{32} & \alpha_{33} \end{bmatrix} \begin{Bmatrix} X_1 \\ X_2 \\ X_3 \end{Bmatrix}.$$

Now, $$\alpha_{11} = \alpha_{12} = \alpha_{21} = \alpha_{13} = \alpha_{31} = \frac{1}{3k},$$

$$\alpha_{22} = \alpha_{32} = \alpha_{33} = \frac{4}{3k},$$

and $$\alpha_{33} = \frac{7}{3k}.$$

Hence, $$\begin{Bmatrix} X_1 \\ X_2 \\ X_3 \end{Bmatrix} = \frac{\omega^2 m}{3k} \begin{bmatrix} 4 & 2 & 1 \\ 4 & 8 & 4 \\ 4 & 8 & 7 \end{bmatrix} \begin{Bmatrix} X_1 \\ X_2 \\ X_3 \end{Bmatrix}.$$

To start the iteration a reasonable guess for the first mode is

$$\begin{Bmatrix} 1 \\ 2 \\ 4 \end{Bmatrix},$$

this is inversely proportional to the mass ratio of the bodies.

Eventually iteration for the first mode gives

$$\begin{Bmatrix} 1.0 \\ 3.2 \\ 4.0 \end{Bmatrix} = \frac{14.4\, m\omega^2}{3k} \begin{Bmatrix} 1.0 \\ 3.18 \\ 4.0 \end{Bmatrix},$$

or $\omega_1 = 0.46\sqrt{k/m}$ rad/s.

To obtain the second principal mode, use the orthogonality relation to remove the first mode from the system matrix:

$$m_1 A_1 A_2 + m_2 B_1 B_2 + m_3 C_1 C_2 = 0.$$

Thus $4m\,(1.0)\,A_2 + 2m\,(3.18)\,B_2 + m\,(4.0)\,C_2 = 0$,

or $A_2 = -1.59\,B_2 - C_2$, since the first mode is $\begin{Bmatrix} 1.0 \\ 3.18 \\ 4.0 \end{Bmatrix}$

Hence, rounding 1.59 up to 1.6,

$$\begin{Bmatrix} A_2 \\ B_2 \\ C_2 \end{Bmatrix} = \begin{bmatrix} 0 & -1.6 & -1 \\ 0 & 1 & 0 \\ 0 & 0 & 1 \end{bmatrix} \begin{Bmatrix} A_2 \\ B_2 \\ C_2 \end{Bmatrix}.$$

When this **sweeping matrix** is combined with the original matrix equation iteration makes convergence to the second mode take place because the first mode is swept out. Thus,

$$\begin{Bmatrix} X_1 \\ X_2 \\ X_3 \end{Bmatrix} = \frac{\omega^2 m}{3k} \begin{bmatrix} 4 & 2 & 1 \\ 4 & 8 & 4 \\ 4 & 8 & 7 \end{bmatrix} \begin{bmatrix} 0 & -1.6 & -1 \\ 0 & 1 & 0 \\ 0 & 0 & 1 \end{bmatrix} \begin{Bmatrix} X_1 \\ X_2 \\ X_3 \end{Bmatrix},$$

$$= \frac{\omega^2 m}{3k} \begin{bmatrix} 0 & -4.4 & -3 \\ 0 & 1.6 & 0 \\ 0 & 1.6 & 3 \end{bmatrix} \begin{Bmatrix} X_1 \\ X_2 \\ X_3 \end{Bmatrix}.$$

Now guess the second mode as $\begin{Bmatrix} 1 \\ 0 \\ -1 \end{Bmatrix}$ and iterate:

$$\begin{Bmatrix} 1 \\ 0 \\ -1 \end{Bmatrix} = \frac{\omega^2 m}{3k} \begin{bmatrix} 0 & -4.4 & -3 \\ 0 & 1.6 & 0 \\ 0 & 1.6 & 3 \end{bmatrix} \begin{Bmatrix} 1 \\ 0 \\ -1 \end{Bmatrix} = \frac{m\omega^2}{k} \begin{Bmatrix} 1 \\ 0 \\ -1 \end{Bmatrix}.$$

Hence $\omega_2 = \sqrt{k/m}$ rad/s, and the second mode was evidently guessed correctly as 1:0:$-$1.

To obtain the third mode, write the orthogonality relation as

$$m_1A_2A_3 + m_2B_2B_3 + m_3C_2C_3 = 0,$$

and $$m_1A_1A_3 + m_2B_1B_3 + m_3C_1C_3 = 0.$$

Substitute $A_1 = 1.0, B_1 = 3.18, C_1 = 4.0,$

and $A_2 = 1.0, B_2 = 0, C_2 = -1.0$, as found above.

Hence $$\begin{Bmatrix} A_3 \\ B_3 \\ C_3 \end{Bmatrix} = \begin{bmatrix} 0 & 0 & 0.25 \\ 0 & 0 & -0.78 \\ 0 & 0 & 1 \end{bmatrix} \begin{Bmatrix} A_3 \\ B_3 \\ C_3 \end{Bmatrix}.$$

When this sweeping matrix is combined with the equation for the second mode the second mode is removed, so that it yields the third mode on iteration;

$$\begin{Bmatrix} X_1 \\ X_2 \\ X_3 \end{Bmatrix} = \frac{\omega^2 m}{3k} \begin{bmatrix} 0 & -4.4 & -3 \\ 0 & 1.6 & 0 \\ 0 & 1.6 & 3 \end{bmatrix} \begin{bmatrix} 0 & 0 & 0.25 \\ 0 & 0 & -0.78 \\ 0 & 0 & 1 \end{bmatrix} \begin{Bmatrix} X_1 \\ X_2 \\ X_3 \end{Bmatrix},$$

$$= \frac{\omega^2 m}{3k} \begin{bmatrix} 0 & 0 & 0.43 \\ 0 & 0 & -1.25 \\ 0 & 0 & 1.75 \end{bmatrix} \begin{Bmatrix} X_1 \\ X_2 \\ X_3 \end{Bmatrix},$$

or $$\begin{Bmatrix} X_1 \\ X_2 \\ X_3 \end{Bmatrix} = 1.75 . \frac{\omega^2 m}{3k} \begin{bmatrix} 0 & 0 & 0.25 \\ 0 & 0 & -0.72 \\ 0 & 0 & 1 \end{bmatrix} \begin{Bmatrix} X_1 \\ X_2 \\ X_3 \end{Bmatrix}.$$

A guess for the third mode shape now has to be made and the iteration procedure carried out once more. In this way the third mode eigenvector is found to be

$$\begin{Bmatrix} 0.25 \\ -0.72 \\ 1.0 \end{Bmatrix},$$

and $$\omega_3 = 1.32 \sqrt{\frac{k}{m}} \text{ rad/s}.$$

The convergence for higher modes becomes more critical if impurities and rounding-off errors are introduced by using the sweeping matrices. It is well to check the highest mode by the inversion of the original matrix equation, which should be equal to the equation formulated in terms of the stiffness influence coefficients.

3.2.2 The Lagrange Equation

Consideration of the energy in a dynamic system together with the use of the Lagrange Equation is a very powerful method of analysis for certain physically complex systems. It is an energy method which allows the equations of motion to be written in terms of any set of generalised coordinates. **Generalised coordinates** are a set of independent parameters which completely specify the system location and which are independent of any constraints. The fundamental form of Lagrange's Equation can be written in terms of the generalised coordinates q_i as follows:

$$\frac{\mathrm{d}}{\mathrm{d}t}\cdot\frac{\partial T}{\partial \dot{q}_i} - \frac{\partial T}{\partial q_i} + \frac{\partial V}{\partial q_i} + \frac{\partial (DE)}{\partial \dot{q}_i} = Q_i,$$

where T is the total kinetic energy of the system, V is the total potential energy of the system, DE is the energy dissipation function when the damping is linear (it is half the rate at which energy is dissipated so that for viscous damping $DE = \frac{1}{2}c\dot{x}^2$), Q_i is a generalised external force (or a non-linear damping force) acting on the system, and q_i is a generalised coordinate that describes the position of the system.

The subscript i denotes n equations for an n degree of freedom system, so that the Lagrange Equation yields as many equations of motion as there are degrees of freedom.

For a free conservative system Q_i and DE are both zero, so that:

$$\frac{\mathrm{d}}{\mathrm{d}t}\cdot\frac{\partial T}{\partial \dot{q}_i} - \frac{\partial T}{\partial q_i} + \frac{\partial V}{\partial q_i} = 0.$$

The full derivation of the Lagrange Equation can be found in *Vibration Theory and Applications* by W. T. Thompson (Allen & Unwin).

Example 11

To isolate a structure from the vibration generated by a machine, the machine is mounted on a large block. The block is supported on springs as shown. Find the equations which describe the motion of the block in the plane of the figure.

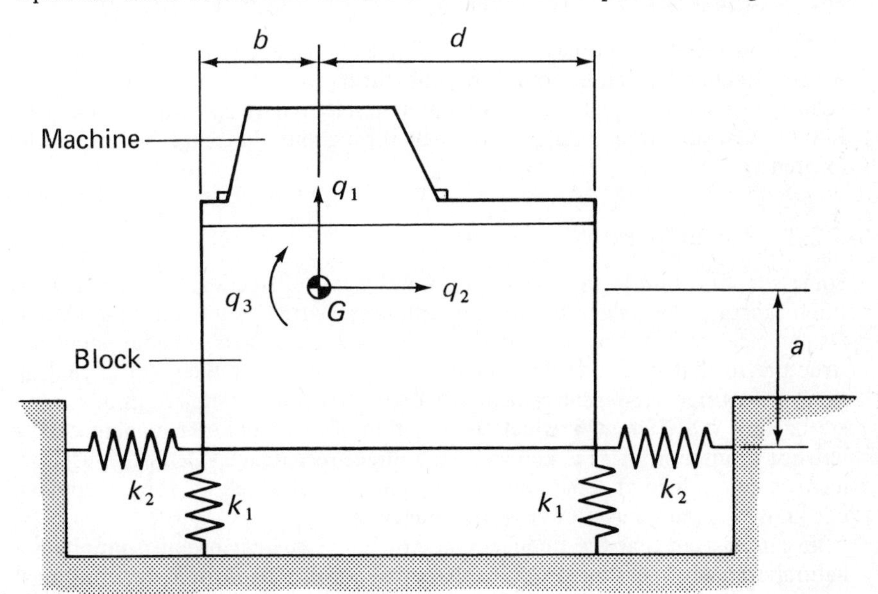

The coordinates used to describe the motion are q_1, q_2 and q_3. These are generalised coordinates because they completely specify the position of the system and are independent of any constraints. If the mass of the block and machine is M, and the total mass moment of inertia about G is I_G, then

$$T = \tfrac{1}{2}M\dot{q}_1^2 + \tfrac{1}{2}M\dot{q}_2^2 + \tfrac{1}{2}I_G\dot{q}_3^2\text{, and}$$

$$V = \text{strain energy stored in the springs,}$$

$$= \tfrac{1}{2}k_1(q_1 + bq_3)^2 + \tfrac{1}{2}k_1(q_1 - dq_3)^2$$

$$+ \tfrac{1}{2}k_2(q_2 - aq_3)^2 + \tfrac{1}{2}k_2(q_2 - aq_3)^2.$$

Now apply the Lagrange Equation with $q_i = q_1$,

$$\frac{\partial T}{\partial q_1} = 0.$$

$$\frac{\partial T}{\partial \dot{q}_1} = M\dot{q}_1\text{, so } \frac{\mathrm{d}}{\mathrm{d}t}\cdot\frac{\partial T}{\partial \dot{q}_1} = M\ddot{q}_1,$$

and
$$\frac{\partial V}{\partial q_1} = k_1(q_1 + bq_3) + k_1(q_1 - dq_3).$$

Thus the first equation of motion is

$$M\ddot{q}_1 + 2k_1q_1 + k_1(b - d)q_3 = 0.$$

Similarly by putting $q_i = q_2$ and $q_i = q_3$, the other equations of motion are obtained as

$$M\ddot{q}_2 + 2k_1q_2 - 2ak_2q_3 = 0,$$

and
$$I_G\ddot{q}_3 + k_1(b - d)q_1 - 2ak_2q_2 + (b^2 + d^2)k_1 + 2a^2k_2\,q_3 = 0.$$

The system therefore has three coordinate-coupled equations of motion. The natural frequencies can be found by substituting $q_i = A_i \sin \omega t$, and solving the resulting frequency equation. It is usually desirable to have all natural frequencies low so that the transmissibility is small throughout the range of frequencies excited.

3.2.3 Receptances

Some simplification in the analysis of multi-degree of freedom undamped structures can often be gained by using receptances, particularly if only the natural frequencies are required. If an *harmonic* force $F \sin \nu t$ acts at some point in a structure so that the structure responds at frequency ν, and the point of application of the force has a displacement $x = X \sin \nu t$, then if the equations of motion are linear $x = \alpha F \sin \nu t$. α, which is a function of the structure parameters and ν but not a function of F, is known as the **direct receptance** at x. If the displacement is determined at some point other than that at which the force is applied, α is known as the transfer or **cross receptance**.

It can be seen that the frequency at which a receptance becomes infinite is a natural frequency of the structure. Receptances can be written for rotational and translational coordinates in a structure, that is the slope and deflection at a point.

Thus, if a body of mass m is subjected to a force $F \sin \nu t$ and the response of the body is $x = X \sin \nu t$,

$$F \sin \nu t = m\ddot{x} = m(-X\nu^2 \sin \nu t) = -m\nu^2 x.$$

Thus $$x = -\frac{1}{m\nu^2} F \sin \nu t,$$

and $$\alpha = -\frac{1}{m\nu^2}.$$ This is the direct receptance of a rigid body.

For a spring, $\alpha = 1/k$. This is the direct receptance of a spring.

In an undamped single degree of freedom model of a structure, the equation of motion is,

$$m\ddot{x} + kx = F \sin \nu t,$$

If $x = X \sin \nu t$, $\alpha = 1/(k - m\nu^2)$. This is the direct receptance of a single degree of freedom structure.

In more complicated structures, it is necessary to be able to distinguish between direct and cross receptances and to specify the points at which the receptances are calculated. This is done by using subscripts. The first subscript indicates the coordinate at which the response is measured and the second indicates that at which the force is applied. Thus α_{pq}, which is a cross receptance, is the response at p divided by the harmonic force applied at q, and α_{pp} and α_{qq} are direct receptances at p and q respectively.

Consider the two degree of freedom system shown in Fig. 3.8.

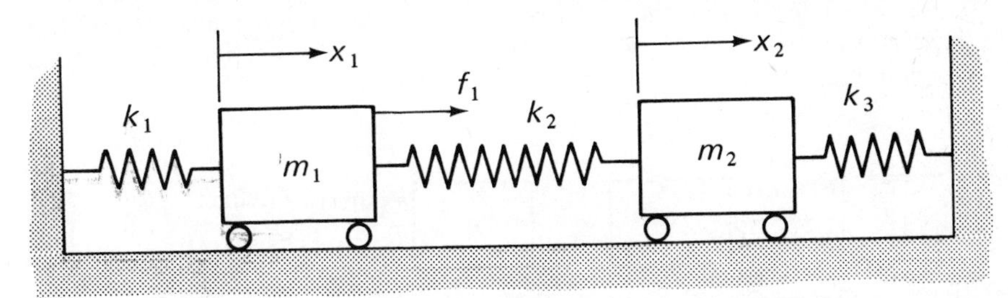

Fig. 3.8 – Two degree of freedom system with forced excitation.

The equations of motion are

$$m_1\ddot{x}_1 + (k_1 + k_2)x_1 - k_2x_2 = f_1,$$

and $$m_2\ddot{x}_2 + (k_2 + k_3)x_2 - k_2x_1 = 0.$$

Let $f_1 = F_1 \sin \nu t$, and assume that $x_1 = X_1 \sin \nu t$ and $x_2 = X_2 \sin \nu t$. Substituting into the equations of motion gives:

$$(k_1 + k_2 - m_1\nu^2)X_1 + (-k_2)X_2 = F_1,$$

and $$(-k_2)X_1 + (k_2 + k_3 - m_2\nu^2)X_2 = 0.$$

Thus $\alpha_{11} = \dfrac{X_1}{F_1} = \dfrac{k_2 + k_3 - m_2\nu^2}{\Delta}$,

where $\Delta = (k_1 + k_2 - m_1\nu^2)(k_2 + k_3 - m_2\nu^2) - k_2^{\,2}$,

α_{11} is a direct receptance, and $\Delta = 0$ is the frequency equation.

Also the cross receptance $\alpha_{21} = \dfrac{X_2}{F_1} = \dfrac{k_2}{\Delta}$.

This system has two more receptances, the responses due to f_2 applied to the second body. Thus α_{12} and α_{22} may be found. It is a fundamental property that $\alpha_{12} \equiv \alpha_{21}$, (Principle of Reciprocity) so that symmetrical matrices result.
A general statement of the system response is

$$X_1 = \alpha_{11}F_1 + \alpha_{12}F_2,$$

and

$$X_2 = \alpha_{21}F_1 + \alpha_{22}F_2,$$

That is,
$$\begin{Bmatrix} X_1 \\ X_2 \end{Bmatrix} = \begin{bmatrix} \alpha_{11} & \alpha_{12} \\ \alpha_{21} & \alpha_{22} \end{bmatrix} \begin{Bmatrix} F_1 \\ F_2 \end{Bmatrix}.$$

Some simplification in the analysis of complicated structures can be achieved by considering the structure to be a number of simple structures, whose receptances are known or easily found, linked together by using the conditions of compatibility and equilibrium. The method is to break the complicated structure model down into sub-structures, and analyse each sub-structure separately. Find each sub-structure receptance at the point where it is connected to the adjacent sub-structure, and 'join' all sub-structures together using the conditions of compatibility and equilibrium.

For example, to find the direct receptance γ_{11} of a dynamic system C at a single coordinate x_1 the system is considered as two sub-systems A and B, as shown in Fig. 3.9.

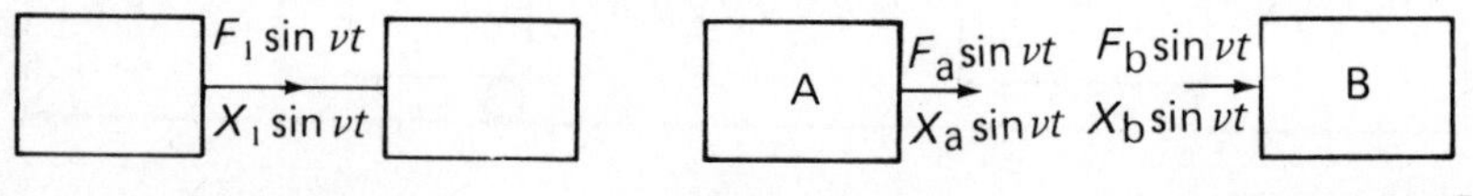

Fig. 3.9 – Dynamic systems.

By definition $\gamma_{11} = \dfrac{X_1}{F_1}$, $\alpha_{11} = \dfrac{X_a}{F_a}$ and $\beta_{11} = \dfrac{X_b}{F_b}$.

Because the systems are connected,

$X_a = X_b = X_1$, (compatibility)

and $F_1 = F_a + F_b$, (equilibrium)

Hence $$\frac{1}{\gamma_{11}} = \frac{1}{\alpha_{11}} + \frac{1}{\beta_{11}},$$

i.e. the system receptance γ can be found from the receptances of the sub-systems.

In a simple spring-body system, sub-systems A and B are the spring and body respectively. Hence $\alpha_{11} = 1/k$ and $\beta_{11} = -1/m\nu^2$ and $1/\gamma_{11} = k - m\nu^2$, as above.

The frequency equation is $\alpha_{11} + \beta_{11} = 0$, because this conditions makes $\gamma_{11} = \infty$.

The receptance technique is particularly useful when it is required to investigate the effects of adding a dynamic system to an existing structure, for example an extra floor, or an air-conditioning plant to a building. Once the receptance of the original structure is known, it is only necessary to analyse the additional system, and then to include this in the original analysis. Furthermore sometimes the receptances of structures and systems are measured, and only available in graphical form.

Some sub-structures, such as those shown in Fig. 3.10, are linked by two co-ordinates, for example deflection and slope at the common point.

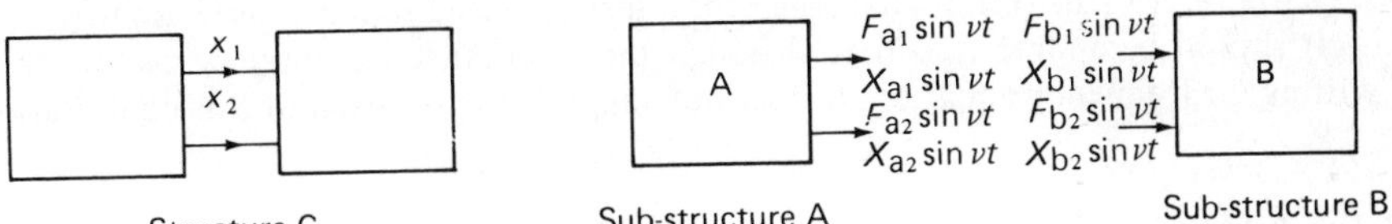

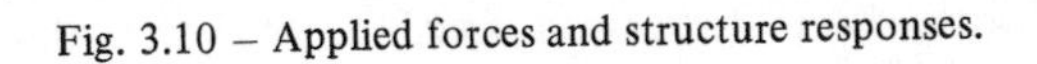
Fig. 3.10 – Applied forces and structure responses.

Now in this case, $X_{a1} = \alpha_{11}F_{a1} + \alpha_{12}F_{a2}$,

$$X_{a2} = \alpha_{21}F_{a1} + \alpha_{22}F_{a2},$$

$$X_{b1} = \beta_{11}F_{b1} + \beta_{12}F_{b2},$$

and $$X_{b2} = \beta_{21}F_{b1} + \beta_{22}F_{b2},$$

The applied forces or moments are $F_1 \sin \nu t$ and $F_2 \sin \nu t$ where

$$F_1 = F_{a1} + F_{b1},$$

and $$F_2 = F_{a2} + F_{b2},$$

Since the sub-structures are linked

$$X_1 = X_{a1} = X_{b1},$$

and $$X_2 = X_{a2} = X_{b2}.$$

Hence if excitation is applied at x_1 only, $F_2 = 0$ and

$$\gamma_{11} = \frac{X_1}{F_1} = \frac{\alpha_{11}(\beta_{11}\beta_{22} - \beta_{12}{}^2) + \beta_{11}(\alpha_{11}\alpha_{22} - \alpha_{12}{}^2)}{\Delta},$$

where $\Delta = (\alpha_{11} + \beta_{11})(\alpha_{22} + \beta_{22}) - (\alpha_{12} + \beta_{12})^2$,

and $$\gamma_{21} = \frac{X_2}{F_1} = \frac{\alpha_{12}(\beta_{11}\beta_{22} - \alpha_{12}\beta_{12}) - \beta_{12}(\alpha_{11}\alpha_{22} - \alpha_{12}\beta_{12})}{\Delta}.$$

If $F_1 = 0$

$$\gamma_{22} = \frac{X_2}{F_2} = \frac{\alpha_{22}(\beta_{11}\beta_{22} - \beta_{12}^2) - \beta_{22}(\alpha_{11}\alpha_{22} - \alpha_{12}^2)}{\Delta}.$$

Since $\Delta = 0$ is the frequency equation the natural frequencies of the structure C are given by

$$\begin{vmatrix} \alpha_{11} + \beta_{11} & \alpha_{12} + \beta_{12} \\ \alpha_{21} + \beta_{21} & \alpha_{22} + \beta_{22} \end{vmatrix} = 0.$$

This is an extremely useful method for finding the frequency equation of a structure because only the receptances of the sub-structures are required. The receptances of many structures and systems have been published in *The Mechanics of Vibration* by R. E. D. Bishop and D. C. Johnson (CUP). By repeated application of this method, a structure can be considered to consist of any number of sub-structures. This technique is therefore, ideally suited to a computer solution.

It should be appreciated that although the receptance technique is useful for writing the frequency equation, it does not simplify the solution of this equation.

3.2.4 Impedance and mobility

Impedance and mobility analysis techniques are frequently applied to systems and structures with many degrees of freedom. However, the method is best introduced by considering simple systems initially.

The **impedance** of a body is the ratio of the amplitude of the *harmonic* exciting force applied, to the amplitude of resulting velocity. The **mobility** is the reciprocal of the impedance. It will be appreciated therefore, that impedance and mobility analysis techniques are similar to those used in the receptance analysis of dynamic systems.

For a body of mass m subjected to an harmonic exciting force represented by $Fe^{j\nu t}$ the resulting motion is $x = Xe^{j\nu t}$.

Thus $Fe^{j\nu t} = m\ddot{x} = -m\nu^2 Xe^{j\nu t}$,

and the receptance of the body, $\dfrac{X}{F} = -\dfrac{1}{m\nu^2}$.

Now $$Fe^{j\nu t} = -m\nu^2 Xe^{j\nu t}$$
$$= mj\nu(j\nu Xe^{j\nu t}) = mj\nu\, v,$$

where v is the velocity of the body, and $v = Ve^{j\nu t}$.

Thus the impedance of a body of mass m is Z_m,

where $Z_m = \dfrac{F}{V} = jm\nu$,

and the mobility of a body of mass m is M_m,

where $$M_m = \frac{V}{F} = \frac{1}{jmv}.$$

Putting $s = jv$ gives

$$Z_m = ms,$$

and $$M_m = \frac{1}{ms},$$

and $$V = sX.$$

For a spring a stiffness k, $Fe^{jvt} = kXe^{jvt}$ and thus $Z_k = F/V = k/s$ and $M_k = s/k$, whereas for a viscous damper of coefficient c, $Z_c = c$ and $M_c = 1/c$.

If these elements of a dynamic structure system are combined so that the velocity is common to all elements, then the impedances may be added to give the system impedance, whereas if the force is common to all elements the mobilities may be added. This is demonstrated below by considering a spring-mass single degree of freedom system with viscous damping, as shown in Fig. 3.11.

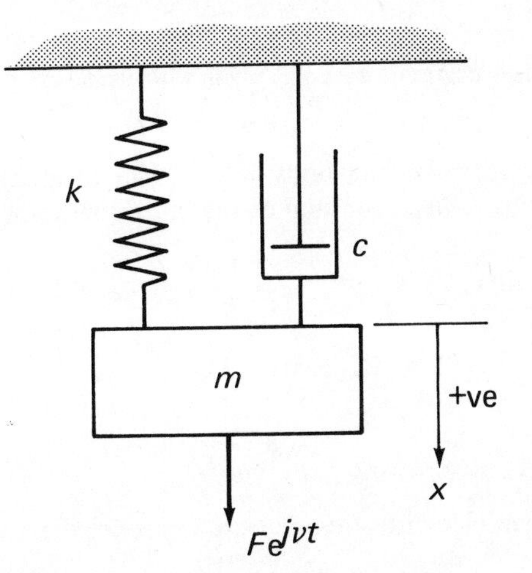

Fig. 3.11 – Single degree of freedom system with elements connected in parallel.

The velocity of the body is common to all elements, so that the force applied is the sum of the forces required for each element. The system impedance,

$$Z = \frac{F}{V} = \frac{F_m + F_k + F_c}{V}$$

$$= Z_m + Z_k + Z_c.$$

Hence $$Z = ms + \frac{k}{s} + c.$$

That is $F = (ms^2 + cs + k)X$,

or $F = (k - mv^2 + jcv)X$.

Hence $X = \dfrac{F}{\sqrt{(k - mv^2)^2 + (cv)^2}}$.

Thus when system elements are connected in parallel their impedances are added to give the system impedance.

In the system shown in Fig. 3.12, however, the force is common to all elements.

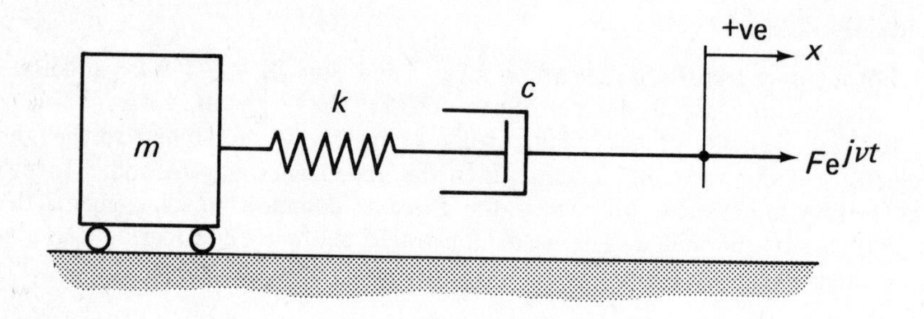

Fig. 3.12 – Single degree of freedom system with elements connected in series.

In this case the force on the body is common to all elements so that the velocity at the driving point is the sum of the individual velocities.

The system mobility, $M = \dfrac{V}{F} = \dfrac{V_m + V_k + V_c}{F}$,

$$= M_m + M_k + M_c,$$

$$= \frac{1}{ms} + \frac{s}{k} + \frac{1}{c}.$$

Thus when system elements are connected in series their mobilities are added to give the system mobility.

In the system shown in Fig. 3.13, the system comprises a spring and damper connected in series with a body connected in parallel.

Thus the spring and damper mobilities can be added, or the reciprocal of their impedances can be added. Hence the system driving point impedance, Z is given by

$$Z = Z_m + \left[\frac{1}{Z_k} + \frac{1}{Z_c}\right]^{-1},$$

$$= ms + \left[\frac{1}{k/s} + \frac{1}{c}\right]^{-1}$$

$$= \frac{mcs^2 + mks + kc}{cs + k}.$$

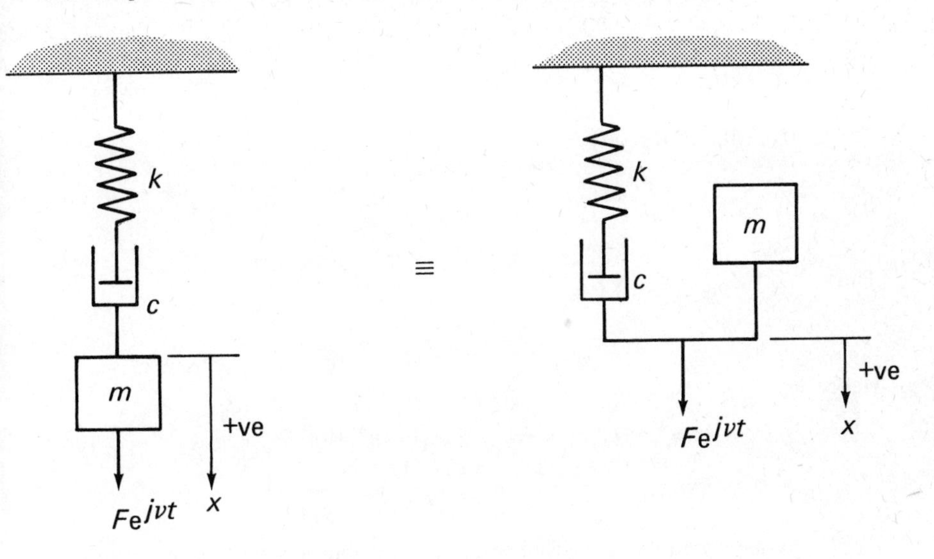

Fig. 3.13 – Single degree of freedom system and impedance analysis model.

Consider the system shown in Fig. 3.14. The spring k_1 and the body m_1 are connected in parallel with each other and are connected in series with the damper c_1.

Thus the driving point impedance Z is

$$Z = Z_{m_2} + Z_{k_2} + Z_{c_2} + Z_1$$

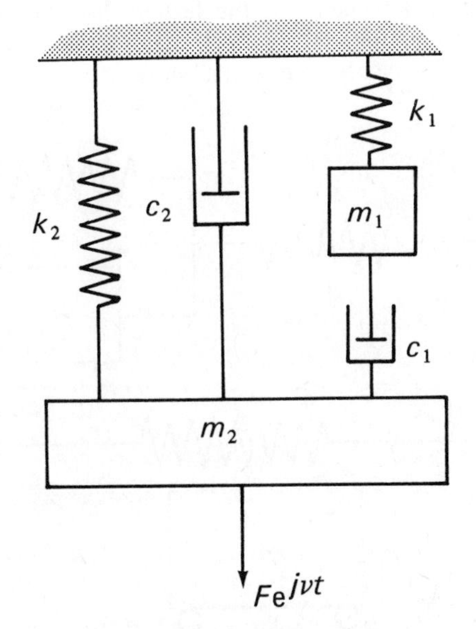

Fig. 3.14 – Dynamic system.

where $Z_1 = \dfrac{1}{M_1}$,

$$M_1 = M_{c_1} + M_2, \; M_2 = \frac{1}{Z_2},$$

and $\quad Z_2 = Z_{k_1} + Z_{m_1}$.

Thus $\quad Z = Z_{m_2} + Z_{k_2} + Z_{c_2} + \dfrac{1}{\dfrac{1}{Z_{c_1}} + \dfrac{1}{Z_{k_1} + Z_{m_1}}}$.

Hence

$$Z = \frac{m_1 m_2 s^4 + (m_1 c_2 + m_2 c_1 + m_1 c_1) s^3 + (m_1 k_1 + m_2 k_2 + c_1 c_2) s^2 + (c_1 k_2 + c_2 k_1 + c_1 k_1) s + k_1 k_2}{s(m_1 s^2 + c_1 s + k_1)}$$

The frequency equation is given when the impedance is made equal to zero or when the mobiiity is infinite. Thus the natural frequencies of the system can be found by putting $s = j\omega$ in the numerator above and setting it equal to zero.

To summarise, the mobility and impedance of individual elements in a dynamic system are calculated on the basis that the velocity is the relative velocity of the two ends of a spring or a damper, but the absolute velocity of the body. Individual impedances are added for elements or sub-systems connected in parallel, and individual mobilities are added for elements or sub-systems connected in series.

Example 12

Find the driving point impedance of the system shown in Fig. 3.6, and hence obtain the frequency equation.

The system of Fig. 3.6 can be re-drawn as shown.

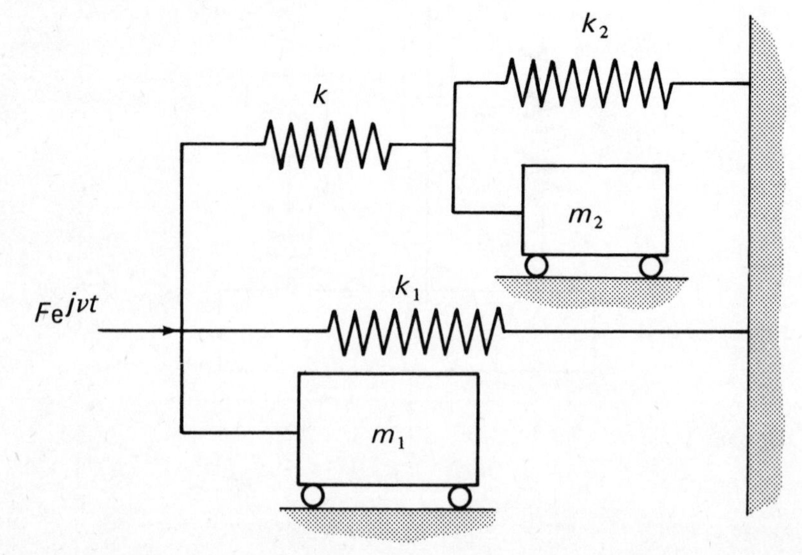

The driving point impedance is therefore

$$Z = Z_{m_1} + Z_{k_1} + \cfrac{1}{\cfrac{1}{Z_k} + \cfrac{1}{Z_{m_2} + Z_{k_2}}},$$

$$= m_1 s + \frac{k_1}{s} + \cfrac{1}{\cfrac{1}{k/s} + \cfrac{1}{m_2 s + k_2/s}},$$

$$= \frac{(m_1 s^2 + k_1)(m_2 s^2 + k + k_2) + (m_2 s^2 + k_2) k}{s(m_2 s^2 + k + k_2)}$$

The frequency equation is obtained by putting $Z = 0$ and $s = j\omega$ thus:

$$(k_1 - m_1\omega^2)(k + k_2 - m_2\omega^2) + k(k_2 - m_2\omega^2) = 0.$$

Chapter 4

The vibration of continuous structures

Continuous structures or structural elements such as beams, rods, and plates can be modelled by discrete mass and stiffness parameters and analysed as multi-degree of freedom systems, but such a model is not sufficiently accurate for most purposes. Furthermore, mass and elasticity cannot always be separated in models of real structures. So mass and elasticity have to be considered as distributed parameters.

For the analysis of structures with distributed mass and elasticity it is necessary to assume a homogeneous, isotropic material which follows Hooke's Law.

Generally, free vibration is the sum of the principal modes. However, in the unlikely event of the elastic curve of the body in which motion is excited coinciding exactly with one of the principal modes, only that mode will be excited. In most continuous structures the rapid damping out of high frequency modes often leads to the fundamental mode predominating.

4.1 LONGITUDINAL VIBRATION OF A THIN UNIFORM BEAM

Consider the longitudinal vibration of a thin uniform beam of cross-sectional area A, material density ρ, and modulus E under an axial force P, as shown in Fig. 4.1.

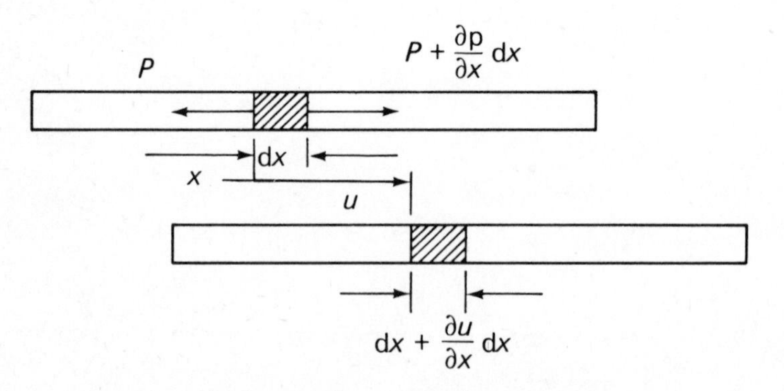

Fig. 4.1 – Longitudinal beam vibration.

The net force acting on the element is $P + \partial P/\partial x \,.\, \mathrm{d}x - P$, and this is equal to the product of the mass of the element and its acceleration.

Hence $$\frac{\partial P}{\partial x} \cdot \mathrm{d}x = \rho A \,\mathrm{d}x \,.\, \frac{\partial^2 u}{\partial t^2}.$$

Now strain $$\frac{\partial u}{\partial x} = \frac{P}{AE}, \text{ so } \frac{\partial P}{\partial x} = AE \frac{\partial^2 u}{\partial x^2}.$$

Thus $$\frac{\partial^2 u}{\partial t^2} = \frac{E}{\rho} \frac{\partial^2 u}{\partial x^2},$$

or $$\frac{\partial^2 u}{\partial x^2} = \frac{1}{c^2} \cdot \frac{\partial^2 u}{\partial t^2}.$$

This is the **Wave Equation.** The velocity of propagation of the displacement or stress wave in the rod is $c = \sqrt{E/\rho}$.

The wave equation can be solved by the method of separation of variables and assuming a solution of the form $u(x, t) = F(x).\, G(t)$.

Substituting this solution into the Wave Equation gives:

$$\frac{\partial^2 F(x)}{\partial x^2} \cdot G(t) = \frac{1}{c^2} \cdot \frac{\partial^2 G(t)}{\partial t^2} \cdot F(x),$$

that is, $$\frac{1}{F(x)} \cdot \frac{\partial^2 F(x)}{\partial x^2} = \frac{1}{c^2} \cdot \frac{1}{G(t)} \cdot \frac{\partial^2 G(t)}{\partial t^2}.$$

The LHS is a function of x only, and the RHS is a function of t only, so partial derivatives are no longer required. Each side must be a constant, $-(\omega/c)^2$ say. (This quantity is chosen for convenience of solution.)

Then $$\frac{\mathrm{d}^2 F(x)}{\mathrm{d}x^2} + \left(\frac{\omega}{c}\right)^2 F(x) = 0,$$

and $$\frac{\mathrm{d}^2 G(t)}{\mathrm{d}t^2} + \omega^2 G(t) = 0.$$

Hence $$F(x) = A \sin \frac{\omega}{c} x + B \cos \frac{\omega}{c} x,$$

and $$G(t) = C \sin \omega t + D \cos \omega t.$$

The constants A and B depend on the boundary conditions, and C and D on the initial conditions. The complete solution to the Wave Equation is therefore:

$$u = \left(A \sin \frac{\omega}{c} x + B \cos \frac{\omega}{c} x\right)(C \sin \omega t + D \cos \omega t).$$

Example 13

Find the natural frequencies and mode shapes of longitudinal vibrations for a free-free beam with initial displacement zero.

Since the beam has free ends, $\partial u/\partial x = 0$ at $x = 0$ and $x = l$.

Now
$$\frac{\partial u}{\partial x} = \left(A\frac{\omega}{c}\cos\frac{\omega}{c}x - B\frac{\omega}{c}\sin\frac{\omega}{c}x\right)(C\sin\omega t + D\cos\omega t).$$

Hence
$$\left(\frac{\partial u}{\partial x}\right)_{x=0} = A\frac{\omega}{c}(C\sin\omega t + D\cos\omega t) = 0, \text{ so that } A = 0,$$

and
$$\left(\frac{\partial u}{\partial x}\right)_{x=l} = \frac{\omega}{c}\left(-B\sin\frac{\omega l}{c}\right)(C\sin\omega t + D\cos\omega t) = 0.$$

Thus $\sin\dfrac{\omega l}{c} = 0$

and
$$\frac{\omega l}{c} = \frac{\omega l}{\sqrt{E/\rho}} = \pi, 2\pi\ldots n\pi\ldots$$

That is
$$\omega_n = \frac{n\pi}{l}\sqrt{\frac{E}{\rho}}, \text{ where } \omega = c/\text{wavelength}.$$

If the initial displacement is zero, $D = 0$ and $u = B\cos n\pi\, x/l.\ \sin n\pi/l.\ \sqrt{E/\rho}\,t$.

4.2 TRANSVERSE VIBRATION OF A THIN UNIFORM BEAM

The transverse or lateral vibration of a thin uniform beam is another vibration problem in which both elasticity and mass are distributed. Consider the moments and forces acting on the element of the beam shown in Fig. 4.2. The beam has a cross-sectional area A, flexural rigidity EI and material of density ρ.

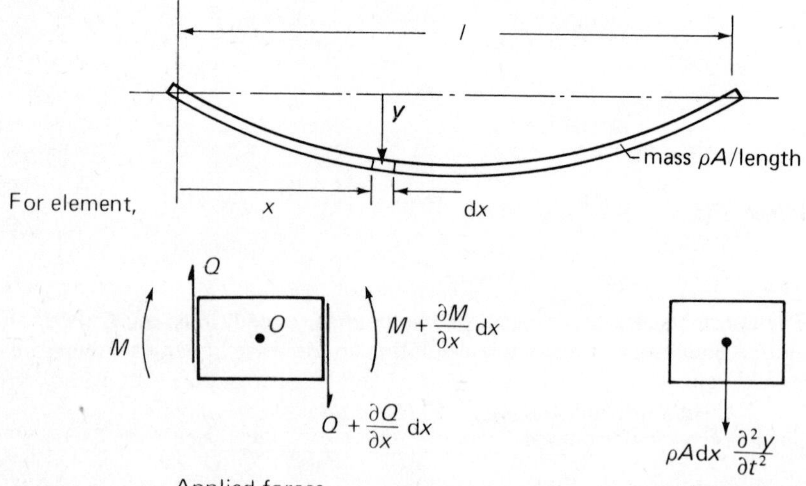

Fig. 4.2 – Transverse beam vibration.

Then for the element, neglecting rotary inertia and shear of the element, taking moments about O gives:

$$M + Q.\frac{dx}{2} + Q.\frac{dx}{2} + \frac{\partial Q}{\partial x}\cdot dx.\frac{dx}{2} = M + \frac{\partial M}{\partial x}\cdot dx.$$

That is, $Q = \dfrac{\partial M}{\partial x}$.

Summing forces in the y direction gives:

$$\frac{\partial Q}{\partial x}\cdot dx = \rho A\ dx.\frac{\partial^2 y}{\partial t^2}.$$

Hence $$\frac{\partial^2 M}{\partial x^2} = \rho A\frac{\partial^2 y}{\partial t^2}.$$

Now EI is a constant for a prismatical beam so

$$M = -EI\frac{d^2 y}{dx^2},\ \text{and}\ \frac{\partial^2 M}{\partial x^2} = -EI\frac{\partial^4 y}{\partial x^4}.$$

Thus $$\frac{\partial^4 y}{\partial x^4} + \frac{\rho A}{EI}\frac{\partial^2 y}{\partial t^2} = 0.$$

This is the general equation for the transverse vibration of a uniform beam.

When a beam performs a normal mode of vibration the deflection at any point of the beam varies harmonically with time, and can be written

$$y = X(B_1 \sin \omega t + B_2 \cos \omega t)$$

where X is a function of x which defines the beam shape of the normal mode of vibration.

Hence $$\frac{d^4 X}{dx^4} = \frac{\rho A}{EI}\omega^2 X = \lambda^4 X,$$

where $\lambda^4 = \rho A \omega^2/EI.$

The general solution to the **beam equation** is

$$X = C_1 \cos \lambda x + C_2 \sin \lambda x + C_3 \cosh \lambda x + C_4 \sinh \lambda x,$$

where the constants $C_{1,2,3,4}$ are determined from the boundary conditions.

For example, consider the transverse vibration of a thin prismatical beam of length l, simply supported at each end. The deflection and bending moment are therefore zero at each end, so that the boundary conditions are $X = 0$ and $d^2X/dx^2 = 0$ at $x = 0$ and $x = l$.

Substituting these boundary conditions into the general solution above gives:

$$x = 0, X = 0;\ \text{thus}\ 0 = C_1 + C_3,$$

and at $$x = 0,\ \frac{d^2 X}{dx^2} = 0;\ \text{thus}\ 0 = C_1 - C_3.$$

That is $C_1 = C_3 = 0$, and $X = C_2 \sin \lambda x + C_4 \sinh \lambda x$.

Now at $x = l, X = 0$, so that $0 = C_2 \sin \lambda l + C_4 \sinh \lambda l$,

and at $x = l, \dfrac{d^2 X}{dx^2} = 0$, so that $0 = C_2 \sin \lambda l - C_4 \sinh \lambda l$.

That is $C_2 \sin \lambda l = C_4 \sinh \lambda l = 0$.

Since $\lambda l \neq 0, \sin \lambda l \neq 0$ and therefore $C_4 = 0$.

Also $C_2 \sin \lambda l = 0$. Since $C_2 \neq 0$, otherwise $X = 0$ for all x, then $\sin \lambda l = 0$.

Hence $X_2 = C_2 \sin \lambda x$, and the solutions to $\sin \lambda l = 0$ give the natural frequencies.

These are $\lambda = 0, \dfrac{\pi}{l}, \dfrac{2\pi}{l}, \dfrac{3\pi}{l} \ldots$

so that $\omega = 0, \left(\dfrac{\pi}{l}\right)^2 \sqrt{\dfrac{EI}{A\rho}}, \left(\dfrac{2\pi}{l}\right)^2 \sqrt{\dfrac{EI}{A\rho}}, \left(\dfrac{3\pi}{l}\right)^2 \sqrt{\dfrac{EI}{A\rho}} \ldots$ rad/s

$\lambda = 0, \omega = 0$ is a trivial solution because the beam is at rest. The lowest or first natural frequency is therefore $\omega_1 = (\pi/l)^2 \sqrt{EI/A\rho}$ rad/s, and the corresponding mode shape is $X = C_2 . \sin \pi x/l$. This is the first mode. $\omega_2 = (2\pi/l)^2 \sqrt{EI/A\rho}$ rad/s is the second natural frequency and the second mode is $X = C_2 \sin 2\pi x/l$, and so on. The mode shapes are drawn in Fig. 4.3.

These sinusoidal vibrations can be superimposed so that any initial conditions can be represented. Other end conditions give frequency equations with the solution

$$\omega = \frac{\alpha}{l^2} \sqrt{\frac{EI}{A\rho}} \text{ rad/s,}$$

where the values of α are given in Table 4.1.

Table 4.1

End conditions	Frequency equation	1st mode	2nd mode	3rd mode	4th mode	5th mode
Clamped–free	$\cos \lambda l \cosh \lambda l = -1$	3.52	22.4	61.7	121.0	199.9
Pinned–pinned	$\sin \lambda l = 0$	9.87	39.5	88.9	157.9	246.8
Clamped–pinned	$\tan \lambda l = \tanh \lambda l$	15.4	50.0	104.0	178.3	272.0
Clamped–clamped or free–free	$\cos \lambda l \cosh \lambda l = 1$	22.4	61.7	121.0	199.9	298.6

The natural frequencies and mode shapes of a wide range of beams and structures are given in *Formulas for Natural Frequency and Mode Shape* by R. D. Blevins (Van Nostrand).

1st mode shape, one half wave:

$$y = C_2 \sin \pi \frac{x}{l} (B_1 \sin \omega_1 t + B_2 \cos \omega_1 t);\ \omega_1 = \left(\frac{\pi}{l}\right)^2 \sqrt{\frac{EI}{A\rho}} \text{ rad/s.}$$

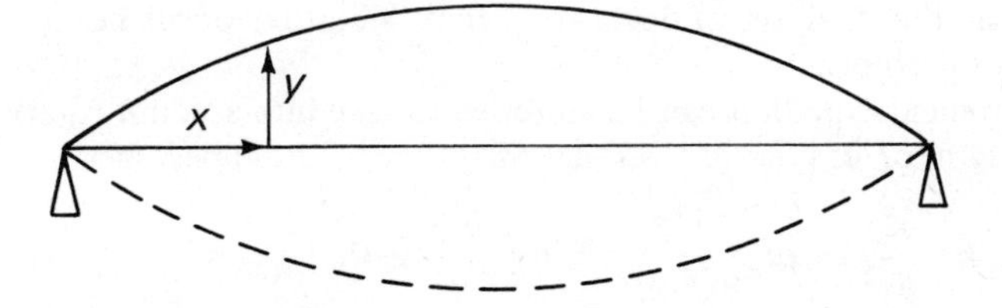

2nd mode shape, two half waves:

$$y = C_2 \sin 2\pi \frac{x}{l} (B_1 \sin \omega_2 t + B_2 \cos \omega_2 t);\ \omega_2 = \left(\frac{2\pi}{l}\right)^2 \sqrt{\frac{EI}{A\rho}} \text{ rad/s.}$$

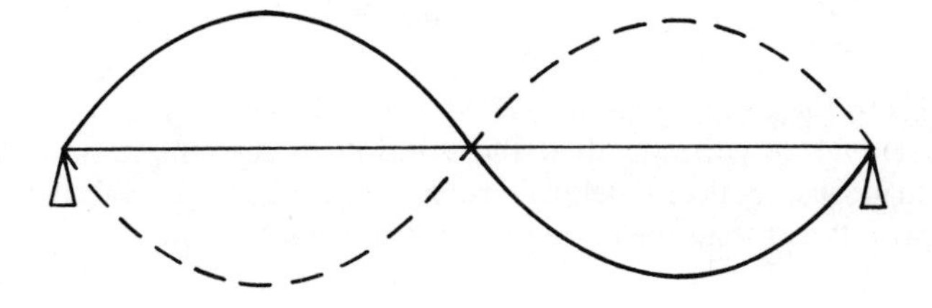

3rd mode shape, three half waves:

$$y = C_2 \sin 3\pi \frac{x}{l} (B_1 \sin \omega_3 t + B_2 \cos \omega_3 t);\ \omega_3 = \left(\frac{3\pi}{l}\right)^2 \sqrt{\frac{EI}{A\rho}} \text{ rad/s.}$$

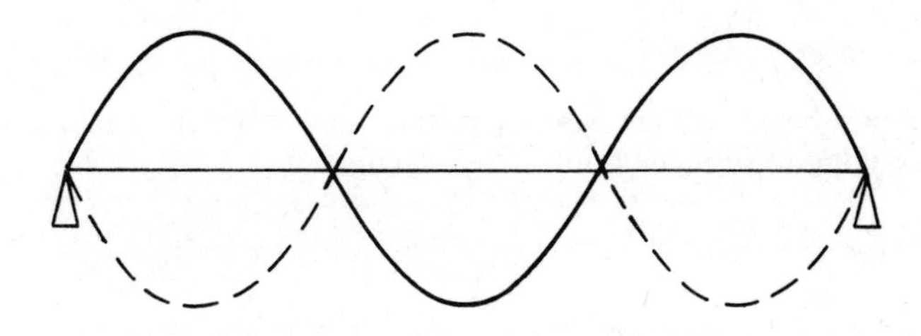

Fig. 4.3 – Transverse beam vibration mode shapes and frequencies.

4.2.1 Rotary inertia and shear effects

When a beam is subjected to lateral vibration so that the depth of the beam is a significant proportion of the distance between two adjacent nodes, rotary inertia of beam elements and transverse shear deformation arising from the severe contortions of the beam during vibration make significant contributions to the lateral deflection. Therefore rotary inertia and shear effects must be taken into account in the analysis of high frequency vibration of all beams, and in all analyses of deep beams.

The moment equation can be modified to take into account rotary inertia by adding a term $\rho I\ \partial^3 y/\partial x\ \partial t^2$, so that the beam equation becomes

$$EI\ \frac{\partial^4 y}{\partial x^4} - \rho I\ \frac{\partial^4 y}{\partial x^2\ \partial t^2} + \rho A\ \frac{\partial^2 y}{\partial t^2} = 0.$$

Shear deformation effects can be included by adding a term

$$\frac{EI\rho}{kg} \cdot \frac{\partial^4 y}{\partial x^2\ \partial t^2}$$

where k is a constant whose value depends on the cross-section of the beam. Generally k is about 0.85. The beam equation then becomes

$$EI\ \frac{\partial^4 y}{\partial x^4} - \frac{EI\rho}{kg} \cdot \frac{\partial^4 y}{\partial x^2\ \partial t^2} + \rho A.\ \frac{\partial^2 y}{\partial t^2} = 0.$$

Solutions to these equations are available, which generally lead to a frequency a few percent more accurate than the solution to the simple beam equation. However, in most cases the modelling errors exceed this. In general, the correction due to shear is larger than the correction due to rotary inertia.

4.2.2 The effect of axial loading

Beams are often subjected to an axial load, and this can have a significant effect on the lateral vibration of the beam. If an axial tension T exists, which is assumed to be constant for small amplitude beam vibrations, the moment equation can be modified by including a term $T.\ \partial^2 y/\partial x^2$, so that the beam equation becomes

$$EI.\ \frac{\partial^4 y}{\partial x^4} - T.\ \frac{\partial^2 y}{\partial x^2} + \rho A.\ \frac{\partial^2 y}{\partial t^2} = 0.$$

Tension in a beam will increase its stiffness and therefore increase its natural frequencies; compression will reduce these quantities.

Example 14

Find the first three natural frequencies of a steel bar 3 cm diameter, which is simply supported at each end, and has a length of 1.5 m. Take $\rho = 7780\ \mathrm{kg/m^3}$ and $E = 208\ \mathrm{GN/m^2}$.

For the bar, $\sqrt{\dfrac{EI}{A\rho}} = \sqrt{\dfrac{208.10^9 .\pi.(0.03)^4/64}{\pi\,(0.03/2)^2\ 7780}}$ m/s^2

$= 38.8$ m/s^2.

Thus $\omega_1 = \dfrac{\pi^2}{1.5^2}\ 38.8 = 170.2$ rad/s and $f_1 = 27.1$ Hz.

Hence $f_2 = 27.1 \times 4 = 108.4$ Hz,

and $f_3 = 27.1 \times 9 = 243.9$ Hz.

If the beam is subjected to an axial tension T, the modified equation of motion leads to the following expression for the natural frequencies;

$$\omega_n{}^2 = \left(\frac{n\pi}{l}\right)^2 \frac{T}{A\rho} + \left(\frac{n\pi}{l}\right)^2 \frac{EI}{A\rho}.$$

For the case when $T = 1000$ N, the correction to ω_1 is

$$\frac{\pi}{1.5}\sqrt{\frac{1000}{\pi\,(0.03/2)^2\ 7780}} = 28.2 \text{ rad/s}.$$

That is $f_1 = 4.5 + 27.1 = 31.6$ Hz.

4.2.3 Transverse vibration of a beam with discrete bodies

In those cases where it is required to find the lowest frequency of transverse vibration of a beam which carries discrete bodies. Dunkerley's method may be used. This is a simple analytical technique which enables a wide range of vibration problems to be solved by using a hand calculator. Dunkerley's method uses the following equation,

$$\frac{1}{\omega_1{}^2} \simeq \frac{1}{P_1{}^2} + \frac{1}{P_2{}^2} + \frac{1}{P_3{}^2} + \frac{1}{P_4{}^2} \ldots,$$

where ω_1 is the lowest natural frequency of a system and $P_1, P_2, P_3 \ldots$ are the frequencies of each body acting alone.

This equation may be obtained for a two degree of freedom system by writing the equations of motion in terms of the influence coefficients as follows:

$$y_1 = \alpha_{11}\, m_1\, \omega^2\, y_1 + \alpha_{12}\, m_2\, \omega^2\, y_2,$$

and $$y_2 = \alpha_{21}\, m_1\, \omega^2\, y_1 + \alpha_{22}\, m_2\, \omega^2\, y_2.$$

The frequency equation is given by;

$$\begin{vmatrix} \alpha_{11}\, m_1\, \omega^2 - 1 & \alpha_{12}\, m_2\, \omega^2 \\ \alpha_{21}\, m_1\, \omega^2 & \alpha_{22}\, m_2\, \omega^2 - 1 \end{vmatrix} = 0.$$

By expanding this determinant, and solving the resulting quadratic equation, it is found that:

$$\omega_{1,2}^2 = \frac{(\alpha_{11}m_1 + \alpha_{22}m_2) \pm \sqrt{(\alpha_{11}m_1 + \alpha_{22}m_2)^2 - 4(\alpha_{11}\alpha_{22} - \alpha_{21}\alpha_{12})}}{2(\alpha_{11}\alpha_{22} - \alpha_{21}\alpha_{12})}$$

Hence it can be shown that:

$$\frac{1}{\omega_1^2} + \frac{1}{\omega_2^2} = \alpha_{11}m_1 + \alpha_{22}m_2 .$$

Now P_1 is the natural frequency of body 1 acting alone,

hence $$P_1^2 = \frac{k_1}{m_1} = \frac{1}{\alpha_{11}m_1} \cdot \text{ Similarly } P_2^2 = \frac{1}{\alpha_{22}m_2} .$$

Thus $$\frac{1}{\omega_1^2} + \frac{1}{\omega_2^2} = \frac{1}{P_1^2} + \frac{1}{P_2^2} .$$

A similar relationship can be derived for systems with more than two degrees of freedom.

If $\omega_2 \gg \omega_1$, the left hand side is approximately $1/\omega_1^2$,

hence $$\frac{1}{\omega_1^2} \simeq \frac{1}{P_1^2} + \frac{1}{P_2^2} .$$

Example 16 shows the application of Dunkerley's method.

4.2.4 Receptance analysis

Many structures can be considered to consist of a number of beams fastened together. Thus if the receptances of each beam are known, the frequency equation of the structure can easily be found by carrying out a sub-structure analysis. (Section 3.2.3). The required receptances can be found by inserting the appropriate boundary conditions in the general solution to the beam equation.

It will be appreciated that this method of analysis is ideal for computer solutions because of its repetitive nature.

For example, consider a beam which is pinned at one end ($x = 0$) and free at the other end ($x = l$). This type of beam is not commonly used in practice, but it is useful for analysis purposes. With an harmonic moment of amplitude M applied to the pinned end, at $x = 0$, $X = 0$ (zero deflection)

and $$\frac{d^2X}{dx^2} = \frac{M}{EI} \qquad \text{(bending moment } M\text{)}$$

and at $x = l$, $$\frac{d^2X}{dx^2} = 0, \qquad \text{(zero bending moment)}$$

and $$\frac{d^3X}{dx^3} = 0 \qquad \text{(zero shear force).}$$

Now, in general $X = C_1 \cos \lambda x + C_2 \sin \lambda x + C_3 \cosh \lambda x + C_4 \sinh \lambda x$.

Thus applying these boundary conditions,

$$0 = C_1 + C_3,$$

and

$$\frac{M}{EI} = -C_1\lambda^2 + C_3\lambda^2.$$

Also

$$0 = -C_1\lambda^2 \cos \lambda l - C_2\lambda^2 \sin \lambda l + C_3\lambda^2 \cosh \lambda l + C_4\lambda^2 \sinh \lambda l,$$

and

$$0 = C_1\lambda^3 \sin \lambda l - C_2\lambda^3 \cos \lambda l + C_3\lambda^3 \sinh \lambda l + C_4\lambda^3 \cosh \lambda l.$$

By solving these equations, $C_{1,2,3,4}$ can be found and substituted into the general solution. It is found that the receptance moment/slope at the pinned end is

$$\frac{(1 + \cos \lambda l . \cosh \lambda l)}{EI\lambda (\cos \lambda l . \sinh \lambda l - \sin \lambda l . \cosh \lambda l)},$$

and at the free end is

$$\frac{2. \cos \lambda l . \cosh \lambda l}{EI\lambda (\cos \lambda l . \sinh \lambda l - \sin \lambda l . \cosh \lambda l)}.$$

The frequency equation is given by

$$\cos \lambda l . \sinh \lambda l - \sin \lambda l . \cosh \lambda l = 0.$$

That is $\tan \lambda l = \tanh \lambda l.$

Moment/deflection receptances can also be found.

By inserting the appropriate boundary conditions into the general solution, the receptance due to an harmonic moment applied at the free end, and harmonic forces applied to either end, can be deduced. Receptances for beams with all end conditions are tabulated in *The Mechanics of Vibration* by R. E. D. Bishop and D. C. Johnson (CUP), thereby greatly increasing the ease of applying this technique.

Example 15

A solar array for a satellite consists of a number of identical solar panels hinged together. The panels are folded up during the launch, and when the satellite is correctly positioned in space the panels are unfolded. The detent mechanism in the hinge, which is a spring loaded pin which locks the array in the desired position, is considered to act as a stiff spring, of torsional stiffness k_T, and the panels are considered to act as beams so that the array, when unfolded, can be modelled by the system shown below.

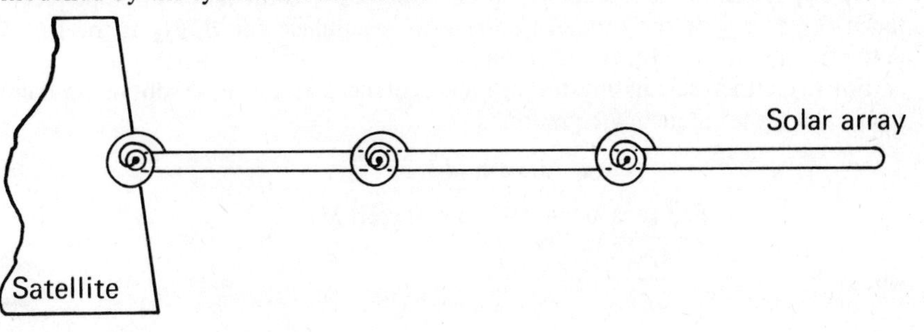

It is required to find the natural frequencies of free vibration of the array, so that excitation of these frequencies, and therefore resonance, can be avoided.

Since all the panels are identical, the receptance technique is relevant for finding the frequency equation. This is because the receptances of each sub-structure are the same, which leads to some simplification in the analysis.

There are two approaches:

(i) To split the array into sub-structures comprising torsional springs and beams.

or (ii) To split the array into sub-structures comprising spring-beam assemblies. This approach results in a smaller number of sub-structures.

Considering the first approach, and only the first element of the array, the sub-structures could be either,

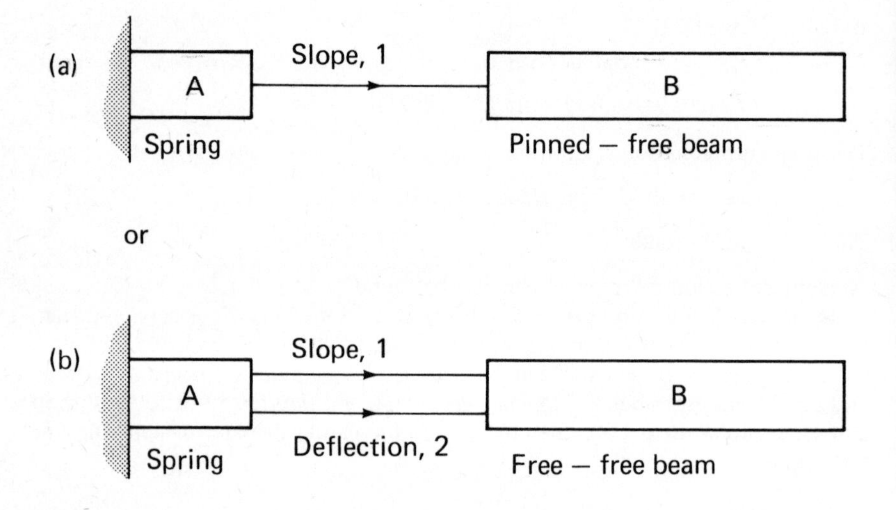

For (a) the frequency equation is $\alpha_{11} + \beta_{11} = 0$, whereas for (b) the frequency equation is

$$\begin{vmatrix} \alpha_{11} + \beta_{11} & \alpha_{12} + \beta_{12} \\ \alpha_{21} + \beta_{21} & \alpha_{22} + \beta_{22} \end{vmatrix} = 0,$$

where α_{11} is the moment/slope receptance for A, β_{11} is the moment/slope receptance for B, β_{12} is the moment/deflection receptance for B, β_{22} is the force/deflection receptance for B and so on.

For (a), either calculating the beam receptances as above, or obtaining them from tables, the frequency equation is

$$\frac{1}{k_T} + \frac{\cos \lambda l . \cosh \lambda l + 1}{EI\lambda (\cos \lambda l \sinh \lambda l - \sin \lambda l \cosh \lambda l)} = 0$$

where $\lambda = \sqrt[4]{\dfrac{A\rho\omega^2}{EI}}$.

For (b), the frequency equation is

$$\begin{vmatrix} \dfrac{1}{k_T} + \dfrac{\cos \lambda l \sinh \lambda l + \sin \lambda l \cosh \lambda l}{EI\lambda (\cos \lambda l \cosh \lambda l - 1)} & \dfrac{-\sin \lambda l . \sinh \lambda l}{EI\lambda^2 (\cos \lambda l \cosh \lambda l - 1)} \\ \dfrac{-\sin \lambda l . \sinh \lambda l}{EI\lambda^2 (\cos \lambda l \cosh \lambda l - 1)} & \dfrac{-(\cos \lambda l \sinh \lambda l - \sin \lambda l \cosh \lambda l)}{EI\lambda^3 (\cos \lambda l \cosh \lambda l - 1)} \end{vmatrix} = 0$$

which reduces to the equation given by method (a).

The frequency equation has to be solved after inserting the structural parameters, to yield the natural frequencies of the structure.

For the whole array it is preferable to use approach (ii), because this results in a smaller number of sub-structures than (i), with a consequent simplification of the frequency equation. However, it will be necessary to calculate the receptances of the spring pinned-free beam if approach (ii) is adopted.

The analysis of structures such as frameworks can also be accomplished by the receptance technique, by dividing the framework to be analysed into beam sub-structures. For example, if the in-plane natural frequencies of a portal frame are required, it can be divided into three sub-structures coupled by the conditions of compatibility and equilibrium, as shown in Fig. 4.4.

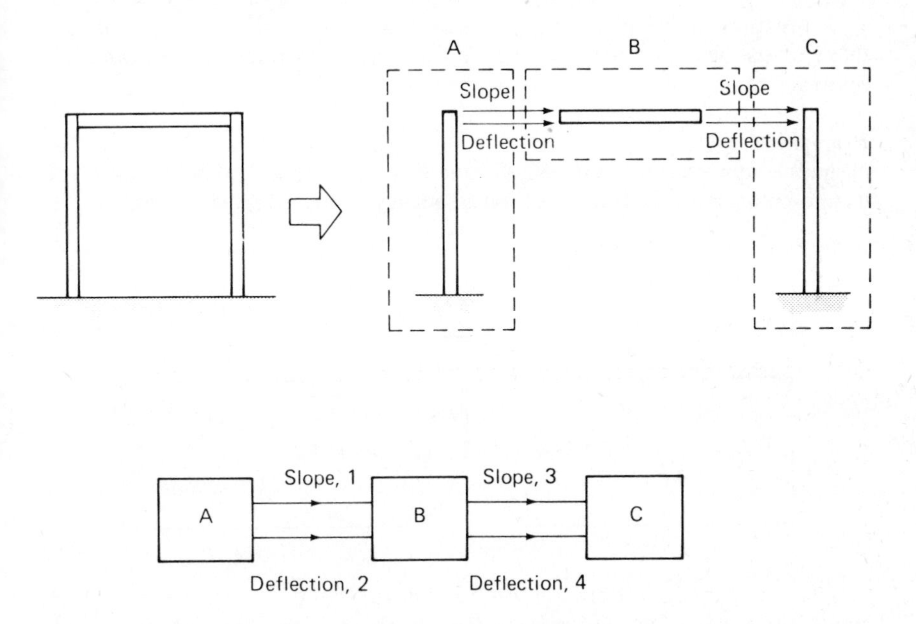

Fig. 4.4 – Portal frame sub-structure analysis.

Sub-structures A and C are cantilever beams undergoing transverse vibration, whereas B is a free-free beam undergoing transverse vibration. Beam B is assumed rigid in the horizontal direction, and the longitudinal deflection of beams A and C is assumed to be negligible.

Because the horizontal member B has no coupling between its horizontal and flexural motion $\beta_{12} = \beta_{14} = \beta_{23} = \beta_{34} = 0$; so that the frequency equation becomes

$$\begin{vmatrix} \alpha_{11}+\beta_{11} & \alpha_{11} & \beta_{13} & 0 \\ \alpha_{21} & \alpha_{22}+\beta_{22} & 0 & \beta_{24} \\ \beta_{31} & 0 & \gamma_{33}+\beta_{33} & \beta_{34} \\ 0 & \beta_{42} & \gamma_{43} & \gamma_{44}+\beta_{44} \end{vmatrix} = 0.$$

4.3 THE ANALYSIS OF CONTINUOUS STRUCTURES BY RAYLEIGH'S ENERGY METHOD

Rayleigh's method, as described in section 2.1.3, gives the lowest natural frequency of transverse beam vibration as

$$\omega^2 = \frac{\int EI\left(\frac{d^2y}{dx^2}\right)^2 dx}{\int y^2\, dm}.$$

A function of x representing y can be determined from the static deflected shape of the beam, or a suitable part sinusoid can be assumed, as shown in the following examples.

Example 16

A simply supported beam of length l and mass m_2 carries a body of mass m_1, at its mid-point. Find the lowest natural frequency of transverse vibration.

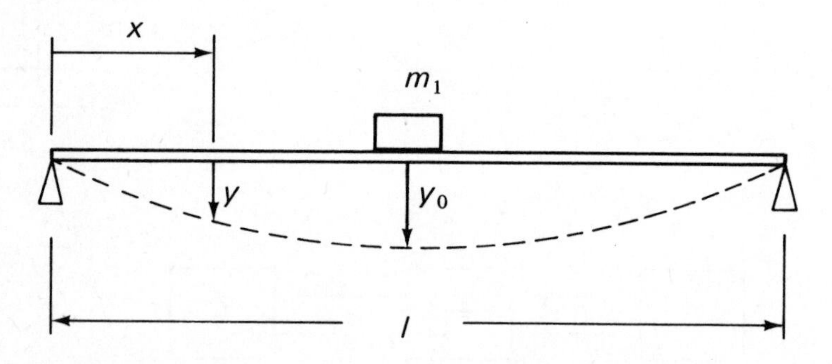

The boundary conditions are $y = 0$ and $d^2y/dx^2 = 0$ at $x = 0$ and $x = l$. These conditions are satisfied by assuming that the shape of the vibrating beam can be represented by a half sine wave. A polynomial expression can be derived for the deflected shape but the sinusoid is usually easier to manipulate.

$y = y_0 . \sin \pi x/l$ is a convenient expression for the beam shape, which agrees with the boundary conditions.

Now $\quad \dot{y} = \dot{y}_0 \sin \frac{\pi x}{l}$ and $\frac{d^2y}{dx^2} = -y_0\left(\frac{\pi}{l}\right)^2 \sin \frac{\pi x}{l}.$

Hence $$\int_0^l EI\left(\frac{d^2y}{dx^2}\right)^2 dx = \int_0^l EI\,y_0^2\left(\frac{\pi}{l}\right)^4 \sin^2\frac{\pi x}{l}\,dx$$
$$= EI\,y_0^2\left(\frac{\pi}{l}\right)^4 \frac{l}{2},$$

and $$\int y^2\,dm = \int_0^l y_0^2 \sin^2\frac{\pi x}{l}\cdot\frac{m_2}{l}\,dx + y_0^2\,m_1$$
$$= y_0^2\left(m_1 + \frac{m_2}{2}\right).$$

Thus $$\omega^2 = \frac{EI\left(\frac{\pi}{l}\right)^4 \frac{l}{2}}{\left(m_1 + \frac{m_2}{2}\right)}.$$

If $$m_2 = 0,\ \omega^2 = \frac{EI}{2}\,\frac{\pi^4}{l^3 m_1} = 48.7\,\frac{EI}{m_1 l^3}.$$

The exact solution is 48 $EI/m_1 l^3$, so the Rayleigh method solution is 1.4% high.

Alternatively the Dunkerley method can be used. Here

$$P_1^2 = \frac{48\,EI}{m_1 l^3} \quad\text{and}\quad P_2^2 = \frac{EI\,\pi^4}{m_2 l^3}.$$

Thus $$\frac{1}{\omega^2} = \frac{m_1 l^3}{48\,EI} + \frac{m_2 l^3}{\pi^4 EI}.$$

Hence $$\omega^2 = \frac{EI\left(\frac{\pi}{l}\right)^4 \frac{l}{2}}{\left(1.015 m_1 + \frac{m_2}{2}\right)},$$

which is very close to the value determined by the Rayleigh method.

Example 17

Find the lowest natural frequency of transverse vibration of a cantilever of mass m, which has a rigid body of mass M attached at its free end.

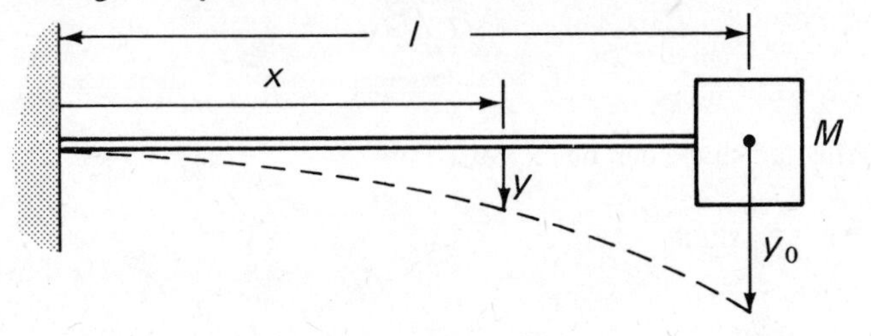

The static deflection curve is $y = (y_0/2l^3)(3lx^2 - x^3)$. Alternatively $y = y_0(1 - \cos \pi x/2l)$ could be assumed.

Hence $$\int_0^l EI\left(\frac{d^2y}{dx^2}\right)^2 dx = EI\int_0^l \left(\frac{y_0}{2l^3}\right)^2 (6l - 6x)^2\, dx = \frac{3EI}{l^3}\, y_0^2,$$

and $$\int y^2\, dm = \int_0^l y^2 \frac{m}{l}\, dx + y_0^2 M$$

$$= \int_0^l \frac{y_0^2}{4l^6} \frac{m}{l} (3lx^2 - x^3)^2\, dx + y_0^2 M$$

$$= y_0^2 \left\{M + \frac{33}{140} m\right\}.$$

Thus $$\omega^2 = \left\{\frac{3\,EI}{\left(M + \dfrac{33}{140} m\right) l^3}\right\},$$

Example 18

A pin ended strut of length l has a vertical axial load P applied. Determine the frequency of free transverse vibration of the strut, and the maximum value of P for stability. The strut has a mass m and a second moment of area I, and is made from material with modulus of elasticity E.

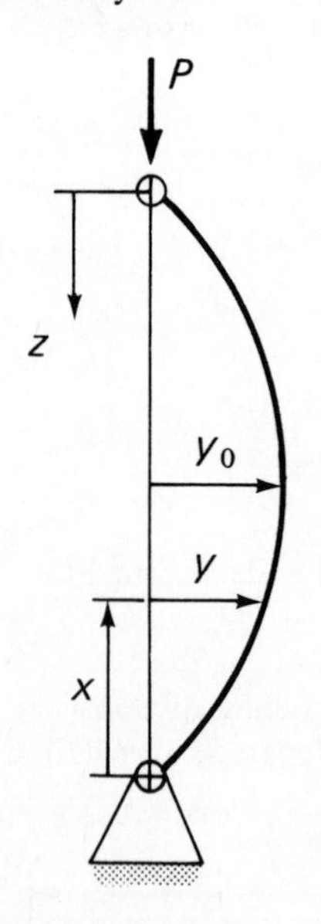

The deflected shape can be expressed by

$$y = y_0 \sin \pi \frac{x}{l},$$

since this function satisfies the boundary conditions of zero deflection and bending moment at $x = 0$ and $x = l$.

Now, $$V_{max} = \tfrac{1}{2}\int EI\left(\frac{d^2y}{dx^2}\right)^2 dx - Pz,$$

where $$\tfrac{1}{2}\int EI\left(\frac{d^2y}{dx^2}\right)^2 dx = \tfrac{1}{2}\int_0^l EI\left(\frac{\pi}{l}\right)^4 . y_0^2 . \sin^2 \pi\frac{x}{l} . dx$$

$$= \frac{EI}{4} . \frac{\pi^4}{l^3} \cdot y_0^2,$$

and $$z = \int_0^l \left(\sqrt{1 + \left(\frac{dy}{dx}\right)^2} - 1\right) dx$$

$$= \int_0^l \tfrac{1}{2}\left(\frac{dy}{dx}\right)^2 dx$$

$$= \tfrac{1}{2}\int_0^l y_0^2 . \left(\frac{\pi}{l}\right)^2 \cos^2 \pi\frac{x}{l} dx$$

$$= \frac{y_0^2}{4} . \frac{\pi^2}{l} .$$

Thus $$V_{max} = \left(\frac{EI}{4} . \frac{\pi^4}{l^3} - \frac{P}{4} . \frac{\pi^2}{l}\right) y_0^2 .$$

Now, $$T_{max} = \tfrac{1}{2}\int y^2 \, dm = \tfrac{1}{2}\int_0^l y^2 . \frac{m}{l} \cdot dx$$

$$= \tfrac{1}{2}\int_0^l y_0^2 \sin^2 \pi\frac{x}{l} . \frac{m}{l} dx = \frac{m}{4} \cdot y_0^2 .$$

Thus $$\omega^2 = \frac{\left(\dfrac{EI}{4} . \dfrac{\pi^4}{l^3} - \dfrac{P}{4} . \dfrac{\pi^2}{l}\right)}{\dfrac{m}{4}},$$

and $$f = \tfrac{1}{2}\sqrt{\frac{EI(\pi/l)^2 - P}{ml}} \text{ Hz.}$$

From section 2.1.4, for stability $\dfrac{dV}{dy_0} = 0$ and $\dfrac{d^2V}{dy_0^2} > 0.$

That is $y_0 = 0$

and $$EI\frac{\pi^2}{l^2} > P.$$

$y_0 = 0$ is the equilibrium position about which vibration occurs, and $P < EI\,\pi^2/l^2$ is the necessary condition for stability. $EI\,\pi^2/l^2$ is known as the Euler buckling load.

4.4 TRANSVERSE VIBRATION OF THIN UNIFORM PLATES

Plates are frequently used as structural elements so that it is sometimes necessary to analyse plate vibration. The analysis considered will be restricted to the vibration of thin uniform flat plates. Non-uniform plates which occur in structures, for example those which are ribbed or bent, may best be analysed by the finite element technique, although exact theory does exist for certain curved plates and shells.

The analysis of plate vibration represents a distinct increase in the complexity of vibration analysis, because it is necessary to consider vibration in two dimensions instead of the single dimension analysis carried out hitherto. It is essentially therefore, an introduction to the analysis of the vibration of multi-dimensional structures.

Consider a thin uniform plate of an elastic, homogeneous isotropic material of thickness h, as shown in Fig. 4.5.

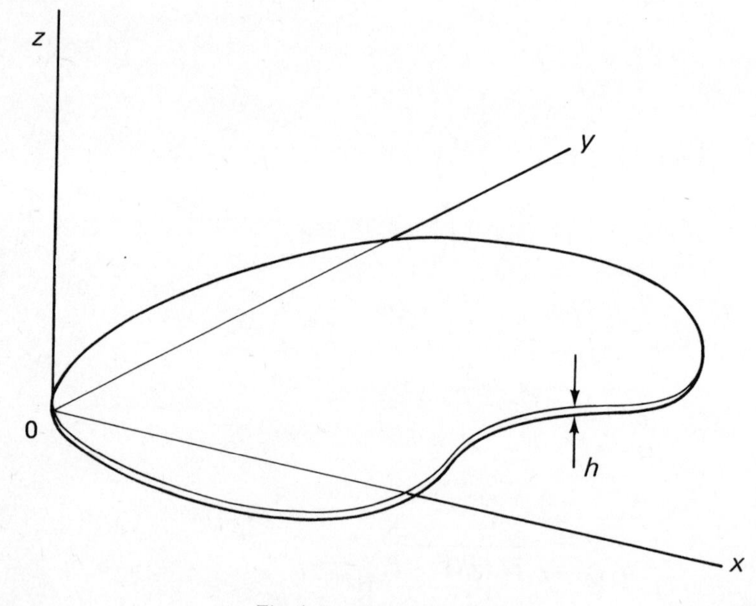

Fig. 4.5 – Thin uniform plate.

If ν is the deflection of the plate at a point (x, y), then it is shown in *Vibration Problems in Engineering* by S. Timoshenko (Van Nostrand), that the potential energy of bending of the plate is

$$\frac{D}{2}\iint\left\{\left(\frac{\partial^2 \nu}{\partial x^2}\right)^2+\left(\frac{\partial^2 \nu}{\partial y^2}\right)^2+2\nu.\frac{\partial^2 \nu}{\partial x^2}\cdot\frac{\partial^2 \nu}{\partial y^2}+2(1-\nu)\left(\frac{\partial^2 \nu}{\partial x.\partial y}\right)^2\right\}\,dx.dy$$

where the flexural rigidity $D = \dfrac{Eh^3}{12(1 - \nu^2)}$,

and ν is Poisson's Ratio.

The kinetic energy of the vibrating plate is

$$\frac{\rho h}{2} \int\int \dot{\nu}^2 \, dx.dy,$$

where ρh is the mass per unit area of the plate.

In the case of a rectangular plate with sides of length a and b, and with simply supported edges, at a natural frequency ω, ν can be represented by

$$\nu = \phi . \sin m\pi \frac{x}{a} . \sin n\pi \frac{y}{b},$$

where ϕ is a function of time.

Thus $$V = \frac{\pi^4 ab}{8} \cdot D. \phi^2 \left(\frac{m^2}{a^2} + \frac{n^2}{b^2}\right)^2,$$

and $$T = \frac{\rho h}{2} \cdot \frac{ab}{4} \cdot \dot{\phi}^2$$

Since $\dfrac{d}{dt}(T + V) = 0$ in a conservative structure,

$$\frac{\rho h}{2} \cdot \frac{ab}{4} \cdot 2\dot{\phi}\ddot{\phi} + \frac{\pi^4 ab}{8} \cdot D.2\phi\dot{\phi} \left(\frac{m^2}{a^2} + \frac{n^2}{b^2}\right)^2 = 0.$$

That is, the equation of motion is

$$\rho h \ddot{\phi} + \pi^4 D \left(\frac{m^2}{a^2} + \frac{n^2}{b^2}\right)^2 \phi = 0.$$

Thus ϕ represents simple harmonic motion and

$$\phi = A \sin \omega_{mn} t + B \cos \omega_{mn} t,$$

where $$\omega_{mn} = \pi^2 \sqrt{\frac{D}{\rho h}} \left(\frac{m^2}{a^2} + \frac{n^2}{b^2}\right) \text{ rad/s.}$$

Now, $$\nu = \phi \sin m\pi \frac{x}{a} \cdot \sin n\pi \frac{y}{b},$$

thus $\nu = 0$ when $\sin m\pi x/a = 0$ or $\sin n\pi y/b = 0$, and hence the plate has nodal lines when vibrating in its normal modes.

Typical nodal lines of the first six modes of vibration of a rectangular plate, simply supported on all edges, are shown in Fig. 4.6.

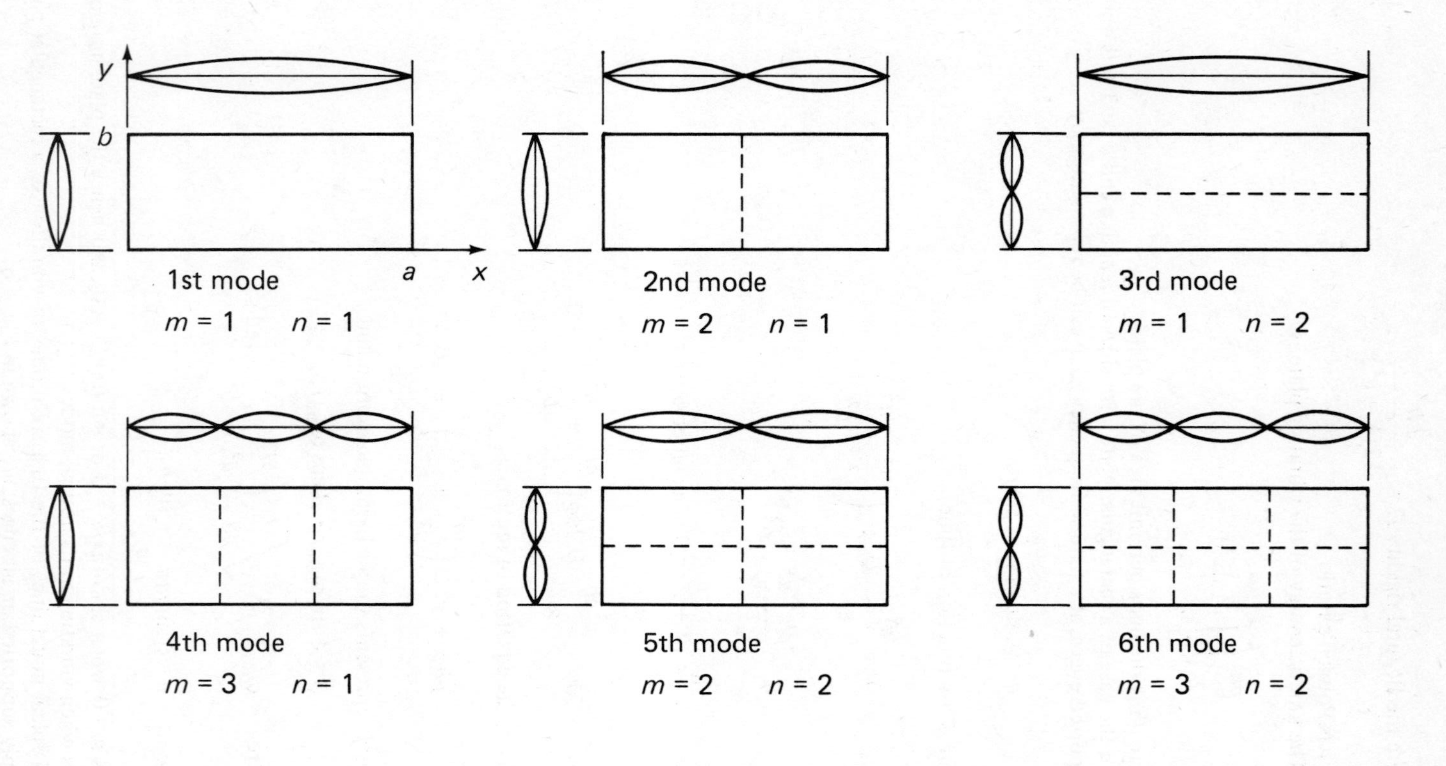

Fig. 4.6 – Transverse plate vibration mode shapes.

An exact solution is only possible using this method if two opposite edges of the plate are simply supported: the other two edges can be free, hinged or clamped. If this is not the case, for example if the plate has all edges clamped, a series solution for ν has to be adopted.

For a simply supported square plate of side $a(= b)$, the frequency of free vibration becomes

$$f = \pi \frac{m^2}{a^2} \sqrt{\frac{D}{\rho h}} \text{ Hz},$$

whereas for a square plate simply supported along two opposite edges and free on the others

$$f = \frac{\alpha}{2\pi a^2} \sqrt{\frac{D}{\rho h}} \text{ Hz},$$

where $\alpha = 9.63$ in the first mode $(1, 1)$, $\alpha = 16.1$ in the second mode $(1, 2)$, and $\alpha = 36.7$ in the third mode $(1, 3)$.

Thus the lowest, or fundamental, natural frequency of a simply supported/free square plate of side l and thickness d is

$$\frac{9.63}{2\pi l^2} \sqrt{\frac{Ed^3}{12(1-\nu^2)\rho d}} = \frac{10.09}{2\pi l^2} \sqrt{\frac{Ed^2}{12\rho}} \text{ Hz},$$

if $\nu = 0.3$.

The theory for beam vibration gives the fundamental natural frequency of a beam simply supported at each end as

$$\frac{1}{2\pi} \left(\frac{\pi}{l}\right)^2 \sqrt{\frac{EI}{A\rho}} \text{ Hz}.$$

If the beam has a rectangular section $b \times d$, $I = \dfrac{bd^3}{12}$ and $A = bd$.

Thus $$f = \frac{1}{2\pi} \left(\frac{\pi}{l}\right)^2 \sqrt{\frac{Ed^2}{12\rho}} \text{ Hz},$$

that is, $$f = \frac{9.86}{2\pi l^2} \sqrt{\frac{Ed^2}{12\rho}} \text{ Hz}.$$

This is very close (within about 2%) to the frequency predicted by the plate theory, although of course beam theory cannot be used to predict all the higher modes of plate vibration, because it assumes that the beam cross section is not distorted. Beam theory becomes more accurate as the aspect ratio of the beam, or plate, increases.

For a circular plate of radius a, clamped at its boundary, it has been shown that the natural frequencies of free vibration are given by

$$f = \frac{\alpha}{2\pi a^2} \sqrt{\frac{D}{\rho h}} \text{ Hz},$$

where α is as given in Table 4.2.

Table 4.2

Number of nodal circles	Number of nodal diameters		
	0	1	2
0	10.21	21.26	34.88
1	39.77	60.82	84.58
2	89.1	120.08	153.81
3	158.18	199.06	242.71

The vibration of a wide range of plate shapes with various types of support is fully discussed in NASA publication SP-160 *Vibration of Plates* by A. W. Leissa.

4.5 THE FINITE ELEMENT METHOD

Many structures, such as a ship hull or engine crankcase, are too complicated to be analysed by classical techniques so that an approximate method has to be used. It can be seen from the receptance analysis of complicated structures that breaking a dynamic structure down into a large number of sub-structures is a useful analytical technique, provided that sufficient computational facilities are available to solve the resulting equations. The finite element method of analysis extends this method to the consideration of continuous structures as a number of elements, connected to each other by conditions of compatibility and equilibrium. Complicated structures can thus be modelled as the aggregate of simpler structures.

The principle advantage of the finite element method is its generality; it can be used to calculate the natural frequencies and mode shapes of any linear elastic structure. However, it is a numerical technique which requires a fairly large computer, and care has to be taken over the sensitivity of the computer output to small changes in input.

For beam type structures the finite element method is similar to the lumped mass method, because the structure is considered to be a number of rigid mass elements of finite size connected by massless springs. The infinite number of degrees of freedom associated with a continuous structure can thereby be reduced to a finite number of degrees of freedom, which can be examined individually.

The finite element method therefore consists of dividing the structure into a series of elements by imaginary lines, and connecting the elements only at the intersections of these lines. These intersections are called **nodes**. It is unfortunate that the word node has been widely accepted for these intersections; this meaning should not be confused with the zero vibration regions referred to in vibration analysis. The stresses and strains in each element are then defined in terms of the displacements and forces at the nodes, and the mass of the elements is lumped at the nodes. A series of equations are thus produced for the displacement of the nodes and hence the structure. By solving these equations the stresses, strains, natural frequencies and mode shapes of the structure can be determined. The accuracy of the finite element method is greatest in the lower modes, and increases as the number of elements in the model increases. The finite element method of analysis is considered in *Techniques of Finite Elements* by B. Irons and S. Ahmad (Ellis Horwood).

Chapter 5

Damping in structures

5.1 SOURCES OF VIBRATION EXCITATION AND ISOLATION

Before attempting to reduce the vibration levels in a structure by increasing its damping, every effort should be made to reduce the vibration excitation at its source. It has to be accepted that many machines and processes generate a disturbing force of one sort or another, but the frequency of the disturbing force should not be at or near a natural frequency of the structure otherwise resonance will occur, with the resulting high amplitudes of vibration and dynamic stresses, and noise and fatigue problems. Resonance may also prevent the structure fulfilling the desired function.

Some reduction in excitation can often be achieved by changing the machinery generating the vibration, but this can usually only be done at the design stage. Resiting equipment may also effect some improvement. However structural vibration caused by external excitation sources such as ground vibration, cross winds or turbulence from adjacent buildings can only be controlled by damping the structure.

In some machines and structures vibrations are deliberately excited as part of the process, for example, in vibratory conveyors and compactors, and in ultrasonic welding. Naturally, nearby structures have to be protected from these vibrations.

Rotating machinery such as fans, turbines, motors and propellers can generate disturbing forces at several different frequencies such as the rotating speed and blade passing frequency. Reciprocating machinery such as compressors and engines can rarely be perfectly balanced, and an exciting force is produced at the rotating speed and at harmonics. Strong vibration excitation in structures can also be caused by pressure fluctuation in gases and liquids flowing in pipes; and intermittent loads such as those imposed by lifts in buildings.

There are two basic types of structural vibration; **steady state vibration** caused by continually running machines such as engines, air-conditioning plants and generators either within the structure or situated in a neighbouring structure, and **transient vibration** caused by a short duration disturbance such as a lorry or train passing over an expansion joint in a road or over a bridge.

Some relief from steady state vibration excitation can often be gained by moving the source of the excitation, since the mass of the vibration generator has some effect on the natural frequencies of the supporting structure. For example, in a building it may be an advantage to move generators and motors to a lower floor, and in a ship re-siting propulsion or service machinery may prove effective. The effect of local stiffening of the structure may prove to be disappointing however because by increasing the stiffness the mass is also increased, so that the change in the $\sqrt{k/m}$ may prove to be very small.

Occasionally a change in the vibration generating equipment can reduce structural vibration levels. For example a change in gear ratios in a mechanical drive system, or a change from a four bladed to a three bladed propeller in a ship propulsion system will alter the excitation frequency provided the speed of rotation is not changed. However, in many cases the running speeds of motors and engines are closely controlled as in electric generators sets, so there is no opportunity for changing the excitation frequency.

If vibration excitation cannot be reduced to acceptable levels, so that the structural response is still too large, some measure of vibration isolation may be necessary.

It is shown in section 2.3.2. that good vibration isolation, that is low force and motion transmissibility, can be achieved by supporting the vibration generator on a flexible low frequency mounting. Thus although disturbing forces are generated, only a small proportion of them are transmitted to the supporting structure. However, this theory assumes that a mode of vibration is excited by an harmonic force passing through the centre of mass of the installation; although this is often a reasonable approximation it rarely actually occurs in practice because, due to a lack of symmetry of the supported machine, several different mountings may be needed to achieve a level installation, and the mass centre is seldom in the same plane as the tops of the mountings. Thus three translational and three rotational modes of vibration have to be considered, and allowance made for coupling between motions in different directions.

Thus the mounting which provides good isolation against a vertical exciting force may allow excessive horizontal motion, because of a frequency component close to the natural frequency of the horizontal mode of vibration. Also a secondary exciting force acting eccentrically from the centre of mass can excite large rotation amplitudes when the frequency is near to that of a rocking mode of an installation.

To limit the motion of a machine installation which generates harmonic forces and moments the mass and inertia of the installation supported by the mountings may have to be increased; that is an inertia block may have to be added to the installation. Example 11 demonstrates the effect of an inertia block, and also how the motions are coupled. If non-metallic mountings are used the dynamic stiffness at the frequencies of interest will have to be found, probably by carrying out further dynamic tests in which the mounting is correctly loaded; they may also possess curious damping characteristics which may be included in the analysis by using the concept of complex stiffness, as discussed in section 2.2.4.

Air bags or bellows are sometimes used for very low frequency mountings where some swaying of the supported system is acceptable. This is an important consideration because if the motion of the inertia block and the machinery is large, pipework and other services may be overstressed which can lead to fatigue failure of these components. Approximate analysis shows that the natural frequency of a body supported on bellows filled with air under pressure is inversely proportional to the square root of the volume of the bellows, so that a change in natural frequency can be effected simply by a change in bellows volume. This can easily be achieved by opening or closing valves connecting the bellows to additional receivers, or by adding a liquid to the bellows. Natural frequencies of 0.5 Hz or even less, are obtainable. An additional advantage of air suspension is that the system can be made self-levelling, when fitted with suitable valves and an air supply. Air pressures of about 5 to 10 times atmospheric pressure are usual.

Greater attenuation of the exciting force at high frequencies can be achieved by using a two-stage mounting. In this arrangement the machine is set on flexible mountings on an inertia block, which is itself supported by flexible mountings. This may not be too expensive to install since in many cases an existing subframe or structure can be used as the inertia block. If a floating floor in a building is used as the inertia block, some allowance must be made for the additional stiffness arising from the air space below it. This can be found by measuring the dynamic stiffness of the floor by means of resonance tests.

Naturally, techniques used for isolating structures from exciting forces arising in machinery and plant can also be used for isolating delicate equipment from vibrations in the structure. For example, sensitive electrical equipment in ships can be isolated from hull vibration, and operating tables and metrology equipment can be isolated from building vibration.

The above isolation systems are all passive; an active isolation system is one in which the exciting force or moment is applied by an externally powered force or couple. The opposing force or moment can be produced by means such as hydraulic rams, out-of-balance rotating bodies or electro-magnetism. Naturally it is essential to have accurate phase and amplitude control, to ensure that the opposing force is always equal, and opposite, to the exciting force. Although active isolation systems can be expensive to install, excellent results are obtainable so that the supporting structure is kept almost completely still. However it must be noted that force actuators such as hydraulic rams must react on another part of the structure.

If, after careful selection and design of machinery and equipment, careful installation and commissioning, and carrying out isolation as necessary the vibration levels in the structure are still too large, then some increase in the damping of the structure is necessary. This is also the case when excitation occurs from sources beyond the designers' control such as cross winds, earthquakes and currents.

Example 19

A machine of mass m generates a disturbing force $F \sin \nu t$; to reduce the force transmitted to the supporting structure, the machine is mounted on a spring of stiffness k with a damper in parallel. Compare the effectiveness of this isolation system for viscous and hysteretic damping.

Viscous damping. From section 2.3.2,

$$T_R = \frac{F_T}{F} = \frac{\sqrt{1 + \left(2\zeta \dfrac{\nu}{\omega}\right)^2}}{\sqrt{\left[1 - \left(\dfrac{\nu}{\omega}\right)^2\right]^2 + \left[2\zeta \dfrac{\nu}{\omega}\right]^2}}$$

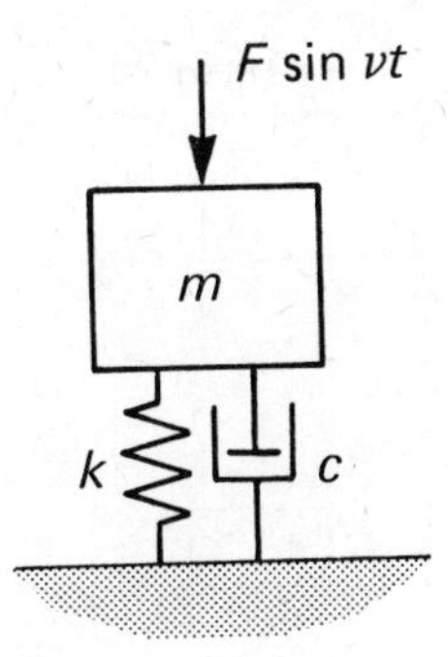

Hysteretic damping. From section 2.2.5,

Putting $\eta = \frac{c\nu}{k} = 2\zeta \frac{\nu}{\omega}$,

$$T_R = \frac{F_T}{F} = \frac{\sqrt{1+\eta^2}}{\sqrt{\left[1-\left(\frac{\nu}{\omega}\right)^2\right]^2+\eta^2}} .$$

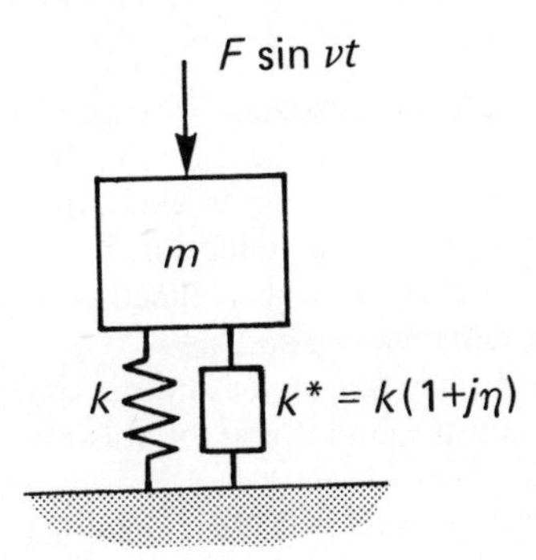

The effectiveness of these isolators can be compared using these expressions for T_R. The results are given in Table 5.1.

It can be seen that the isolation effects are similar for the viscous and hysteretically damped isolators, except at high frequency ratios when the hysteretic damping gives much better attenuation of T_R. At these frequencies it is better to decouple the viscous damped isolator by attaching small springs or rubber bushes at each end.

Table 5.1

	Viscously damped isolator	Hysteretically damped isolator
Value of T_R when $\nu = 0$	1	1
Frequency ratio ν/ω for resonance	1	1
Value of T_R at resonance	$\dfrac{\sqrt{1+(2\zeta)^2}}{2\zeta} \simeq \dfrac{1}{2\zeta}$	$\dfrac{\sqrt{1+\eta^2}}{\eta} \simeq \dfrac{1}{\eta}$
Value of T_R when $\nu/\omega = \sqrt{2}$	1	1
Frequency ratio ν/ω for isolation	$>\sqrt{2}$	$>\sqrt{2}$
High frequency, $\nu/\omega \gg 1$, attenuation of T_R	$\dfrac{2\zeta}{\nu/\omega}$	$\dfrac{1}{(\nu/\omega)^2}$

5.2 EFFECTS OF DAMPING ON VIBRATION RESPONSE OF STRUCTURES

It is desirable for all structures to possess sufficient damping so that their response to the expected excitation is acceptable. Increasing the damping in a structure will reduce its response to a given excitation. Thus if the damping in a structure is increased there will be a reduction in vibration and noise, and the dynamic stresses in the structure will be reduced with a resulting benefit to the fatigue life. Naturally the converse is also true.

However it should be noted that increasing the damping in a structure is not always easy, it can be expensive, and it may be wasteful of energy during normal operating conditions.

Some structures need to possess sufficient damping so that their response to internally generated excitation is controlled: for example a crane structure has to have a heavily damped response to sudden loads, and machine tools must have adequate damping so that a heavily damped response to internal excitation occurs, so that the cutting tool produces a good and accurate surface finish with a high cutting speed. Other structures such as chimneys and bridges must possess sufficient damping so that their response to external excitation such as cross winds does not produce dynamic stresses likely to cause failure through fatigue. In motor vehicles, buildings and ships, noise and vibration transmission through an inadequately damped structure may be a major consideration.

Before considering methods for increasing the damping in a structure, it is necessary to be able to measure structural damping accurately.

5.3 THE MEASUREMENT OF STRUCTURAL DAMPING

It must be appreciated that in any structure a number of mechanisms contribute to the total damping. Different mechanisms may be significant at different stress levels, temperatures or frequencies. Thus damping is both frequency and mode dependent, both as to its mechanism and its magnitude. In discussing the effect of various variables on the total damping in a structure it is essential therefore to define all the operating conditions.

Sometimes it is not possible to measure the damping occurring in a structure on its own. For example ships have to be tested in water, which significantly effects the total damping. However since the ship always operates in water, this total damping is relevant; what is not clear is how changes in the structure will effect the total in-water damping; cargos may also have some effect. On the other hand, a structure such as a machine tool can be tested free of liquids and workpiece; indeed the damping of each structural component can be measured in an attempt to find the most significant source of damping, and hence the most efficient way of increasing the total damping in the structure.

In all cases when damping measurements are being carried out a clear idea of exactly what is being measured is essential. It must be noted that in some tests carried out the damping within the test system itself has, unfortunately, been the major contributor to the total damping.

It has been seen in chapter 2 that the free-decay method is a convenient way for assessing the damping in a structure. The structure is set into free vibration by a shock load such as a small explosive; the fundamental mode dominates the response since all the higher modes are damped out quite quickly. It is not usually possible to excite any mode other than the fundamental using this method. By measuring and recording the decay in the oscillation the logarithmic decrement Λ is found, where

$$\Lambda = \ln\left(\frac{\text{Amplitude of motion}}{\text{Amplitude of motion one cycle later}}\right)$$

If the damping is viscous, or acts in an equivalent viscous manner Λ will be a constant irrespective of the amplitude. To check this, the natural logarithm of the amplitudes can be plotted against cycles of motion; viscous damping gives a straight line, as shown in Fig. 5.1.

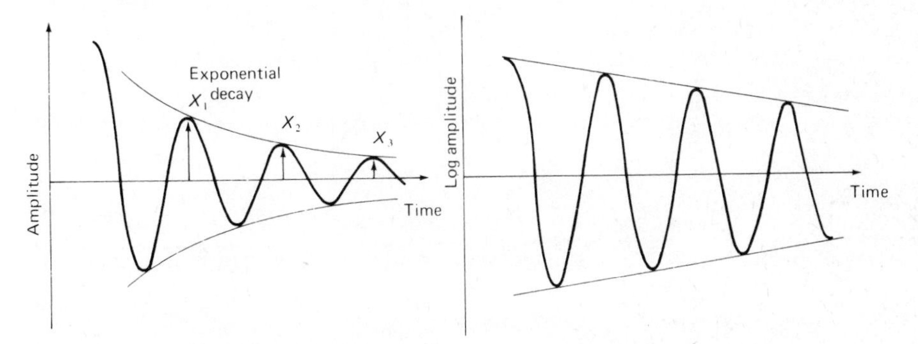

Fig. 5.1 – Vibration decay for viscous damped system.

For viscous damping,

$$\Lambda = \ln\left(\frac{X_1}{X_2}\right) = \ln\left(\frac{X_2}{X_3}\right) = \ln\left(\frac{X_3}{X_4}\right) = \ldots = \ln\left(\frac{X_{n-1}}{X_n}\right).$$

Thus $$n\Lambda = \ln \frac{X_1}{X_2} \cdot \frac{X_2}{X_3} \cdot \frac{X_3}{X_4} \cdots \frac{X_{n-2}}{X_{n-1}} \cdot \frac{X_{n-1}}{X_n},$$

that is $$\Lambda = \frac{1}{n} \cdot \ln\left(\frac{X_1}{X_n}\right),$$

which is a useful expression to use if Λ is small. Note that $\Lambda = 2\pi\zeta/\sqrt{1-\zeta^2}$, and for low damping $\Lambda \simeq 2\pi\zeta$.

There are several ways of expressing the damping in a structure, one of the most common is by the Q factor.

When a structure is forced into resonance by an harmonic exciting force, the ratio of the maximum dynamic displacement at steady state conditions to the static displacement under a similar force is called the Q factor.

That is $$Q = \frac{X_{\text{max.dyn.}}}{X_{\text{static}}} = \frac{1}{2\zeta} \quad \text{(section 2.3.1)}.$$

Since a structure can be excited into resonance at any of its modes, a Q factor can be determined for each mode. This can be done by carrying out a resonance test; Fig. 5.2 shows a typical response.

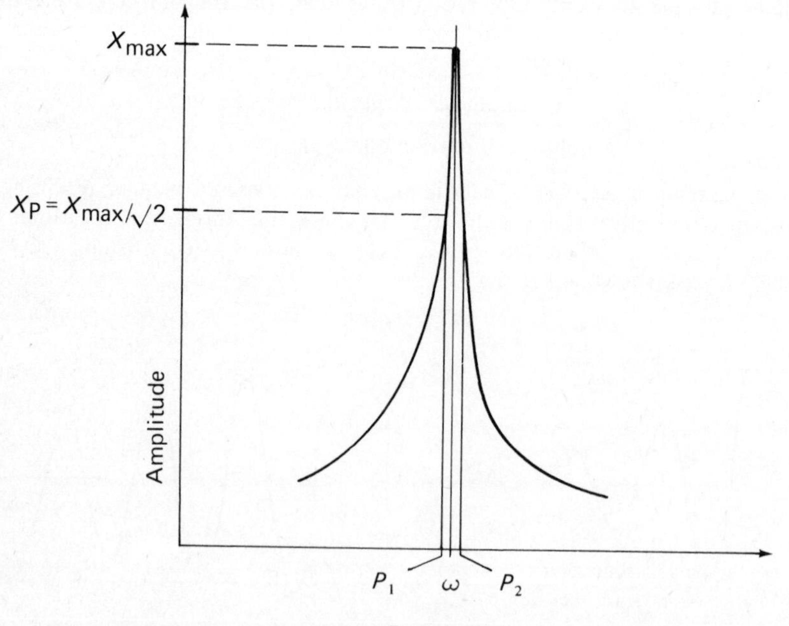

Fig. 5.2 – Amplitude-frequency response, single resonance.

If X_{static} cannot be determined, the Q factor can be found by using the half power point method. This method requires very accurate measurement of the vibration amplitude for excitation frequencies in the region of resonance.

If the damping is assumed to be light, which is the case for most structures, $\omega = \sqrt{k_{mode}/m_{mode}}$, where k_{mode} and m_{mode} are the effective stiffness and mass respectively of the structure for the mode of vibration considered. Once X_{max} and ω have been located, the so called half-power points are found when the amplitude is $X_p = X_{max}/\sqrt{2}$ and the corresponding frequencies either side of ω, p_1 and p_2 determined. Since the energy dissipated per cycle is proportional to X^2, the energy dissipated is reduced by 50% when the amplitude is reduced by a factor $1/\sqrt{2}$.

Now,
$$X = \frac{F/k}{\sqrt{\left[1-\left(\frac{\nu}{\omega}\right)^2\right]^2 + \left[2\zeta\frac{\nu}{\omega}\right]^2}}.$$

Thus
$$X_{max} = \frac{F/k}{2\zeta} \quad \left(\zeta \text{ small, so } X_{max} \text{ occurs at } \frac{\nu}{\omega} = 1\right),$$

and
$$X_p = \frac{X_{max}}{\sqrt{2}} = \frac{F/k}{\sqrt{2}.\,2\zeta} = \frac{F/k}{\sqrt{\left[1-\left(\frac{p}{\omega}\right)^2\right]^2 + \left[2\zeta\frac{p}{\omega}\right]^2}}.$$

Hence
$$\left[1-\left(\frac{p}{\omega}\right)^2\right]^2 + \left[2\zeta\frac{p}{\omega}\right]^2 = 8\zeta^2,$$

and
$$\left(\frac{p}{\omega}\right)^2 = (1-2\zeta^2) \pm 2\zeta\sqrt{1-\zeta^2}.$$

That is
$$\frac{p_2^2 - p_1^2}{\omega^2} = 4\zeta\sqrt{1-\zeta^2} \simeq 4\zeta, \text{ if } \zeta \text{ is small.}$$

Since
$$\frac{p_2^2 - p_1^2}{\omega^2} = \left(\frac{p_2 - p_1}{\omega}\right)\left(\frac{p_2 + p_1}{\omega}\right) = 2\left(\frac{p_2 - p_1}{\omega}\right),$$

because $(p_1 + p_2)/\omega = 2$, that is a symmetrical response curve is assumed for small ζ.

Thus
$$\frac{p_2 - p_1}{\omega} = 2\zeta = \frac{\Delta\omega}{\omega} = \frac{1}{Q},$$

where $\Delta\omega$ is the frequency bandwidth at the half power points.

Thus, for light damping, the damping ratio ζ and hence the Q factor associated with any mode of structural vibration can be found from the amplitude-frequency measurements at resonance and the half power points. Care is needed to ensure that the exciting device does not load the structure and alter the frequency response and the damping, and also that the neighbouring modes do not affect the purity of the mode whose resonance response is being measured. Some difficulty is often encountered in measuring X_{max} accurately. Fig. 5.3

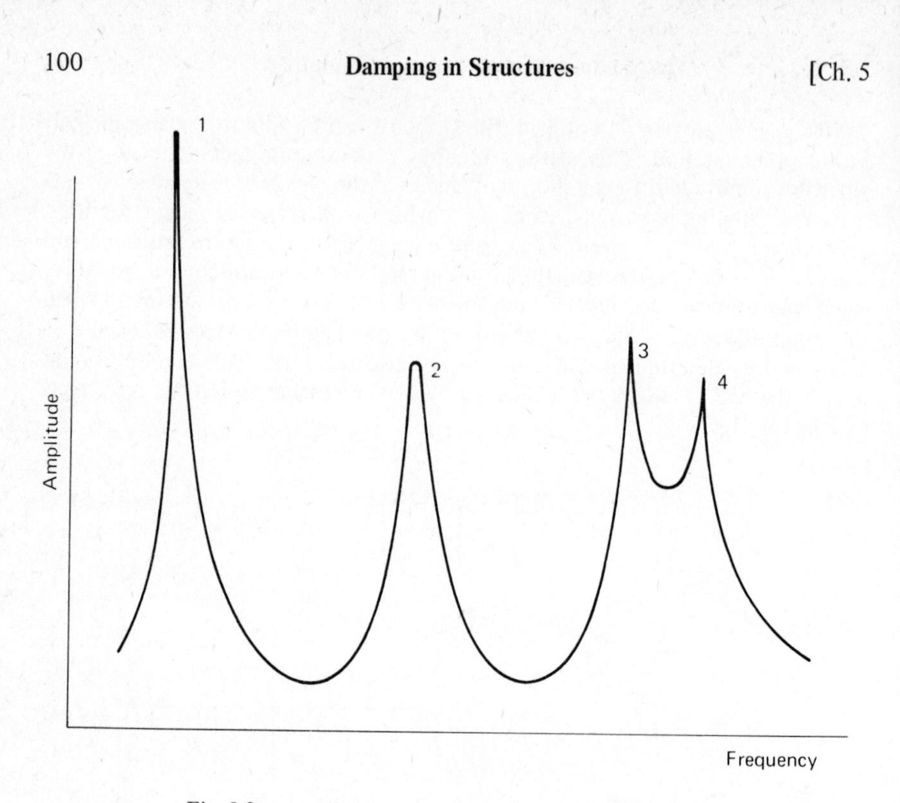

Fig. 5.3 – Amplitude-frequency response, multi resonance.

shows a response in which mode 1 is difficult to measure accurately because of the low damping, that is, the high Q factor. It is difficult to assess the peak amplitude and hence there may be significant errors in the half power points location, and a large % error in $\Delta\omega$ because it is so small. Measurements for mode 2 would probably give an acceptable value for the Q factor for this mode, but modes 3 and 4 are so close together that they interfere with each other, and the half power points cannot be accurately found from this data. Some typical values of the frequency bandwidth at the half power points are given in Table 5.2.

In real structures a very high Q at a low frequency, or a very low Q at a high frequency seldom occurs, but it can be appreciated from the above that very real measuring difficulties can be encountered when trying to measure bandwidths of only a few Hz accurately, even if the amplitude of vibration can be determined.

Table 5.2

Q factor	500	50	5
Resonance frequency (Hz)	Frequency bandwidth (Hz)		
10	0.02	0.2	2
100	0.2	2	20
1000	2	20	200

An improvement in accuracy can often be obtained by measuring both amplitude and phase of the response for a range of exciting frequencies. Consider a single degree of freedom system under forced excitation $Fe^{j\nu t}$. The equation of motion is

$$m\ddot{x} + c\dot{x} + kx = Fe^{j\nu t}.$$

A solution $x = Xe^{j\nu t}$ can be assumed, so that

$$-m\nu^2 X + jc\nu X + kX = F.$$

Hence the receptance $\dfrac{X}{F} = \dfrac{1}{(k - m\nu^2) + jc\nu}$,

$$= \frac{k - m\nu^2}{(k - m\nu^2)^2 + (c\nu)^2} - j\frac{c\nu}{(k - m\nu^2)^2 + (c\nu)^2}.$$

That is, X/F is complex receptance with two vectors $\mathrm{Re}(X/F)$ in phase with the force, and $\mathrm{Im}(X/F)$ in quadrature with the force. The locus of the end point of receptance vector X/F as ν varies is shown for a given value of c in Fig. 5.4. This is obtained by calculating Real and Imaginary components of X/F for a range of frequencies.

Experimentally this curve can be obtained by plotting the measured amplitude and phase of (X/F) for each exciting frequency.

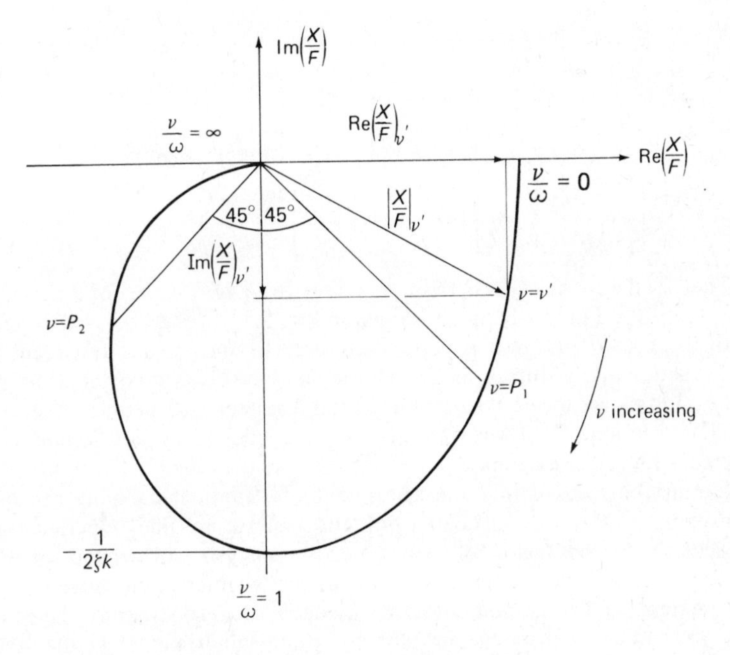

Fig. 5.4 – Receptance vector locus for system with viscous damping.

Since $$\tan\phi = \frac{k - m\nu^2}{c\nu},$$

$$1 = \frac{k - mp_1^2}{cp_1} \quad \text{and} \quad -1 = \frac{k - mp_2^2}{cp_2},$$

Thus $$mp_1^2 + cp_1 - k = 0 \quad \text{and} \quad mp_2^2 - cp_2 - k = 0$$

Subtracting one equation from the other, gives $p_2 - p_1 = c/m$, or

$$\frac{p_2 - p_1}{\omega} = \frac{\Delta\omega}{\omega} = 2\zeta = \frac{1}{Q}.$$

That is, the receptance at resonance lies along the imaginary axis, and the half power points occur when $\phi = 45^\circ$ and 135°. If experimental results are plotted on these axes a smooth curve can be drawn through them so that the half power points can be accurately located. Interference from adjacent modes can thus be reduced.

The method is even more effective when the damping is hysteretic, because in this case the receptance is, from section (2.3.4)

$$\frac{X}{F} = \frac{1}{(k - m\nu^2) + j\eta k},$$

so that $$\mathrm{Re}\left(\frac{X}{F}\right) = \frac{k - m\nu^2}{(k - m\nu^2)^2 + (\eta k)^2},$$

and $$\mathrm{Im}\left(\frac{X}{F}\right) = \frac{-\eta k}{(k - m\nu^2)^2 + (\eta k)^2}.$$

Thus $$\left[\mathrm{Re}\left(\frac{X}{F}\right)\right]^2 + \left[\mathrm{Im}\left(\frac{X}{F}\right)\right]^2 = \frac{1}{(k - m\nu^2)^2 + (\eta k)^2},$$

or $$\left[\mathrm{Re}\left(\frac{X}{F}\right)\right]^2 + \left[\mathrm{Im}\left(\frac{X}{F}\right) - \frac{1}{2\eta k}\right]^2 = \frac{1}{2\eta k}^2.$$

That is, the locus of (X/F) as ν increases from zero is part of a circle, centre $(0, -1/2\eta k)$ and radius $1/2\eta k$, as shown in Fig. 5.5.

In this case therefore it is particularly easy to draw an accurate locus from a few experimental results, even though they may be distorted by adjacent modes, and p_1 and p_2 are located on the horizontal diameter of the circle.

This technique is known variously as a frequency locus plot, Kennedy–Pancu diagram or Nyquist diagram.

It must be realised that the assessment of structural damping can only be approximate. It is very difficult to obtain accurate, reliable, experimental data particularly in the region of resonance; the analysis will depend on whether viscous or hysteretic damping is assumed, modal interaction usually occurs to some extent, and some non-linearity may occur in a real structure. These effects may cause the frequency-locus plot to rotate and translate in the Re(X/F), Im(X/F) plane. In these cases the resonance frequency can be found from that part of the plot where the greatest rate of change of phase with frequency occurs.

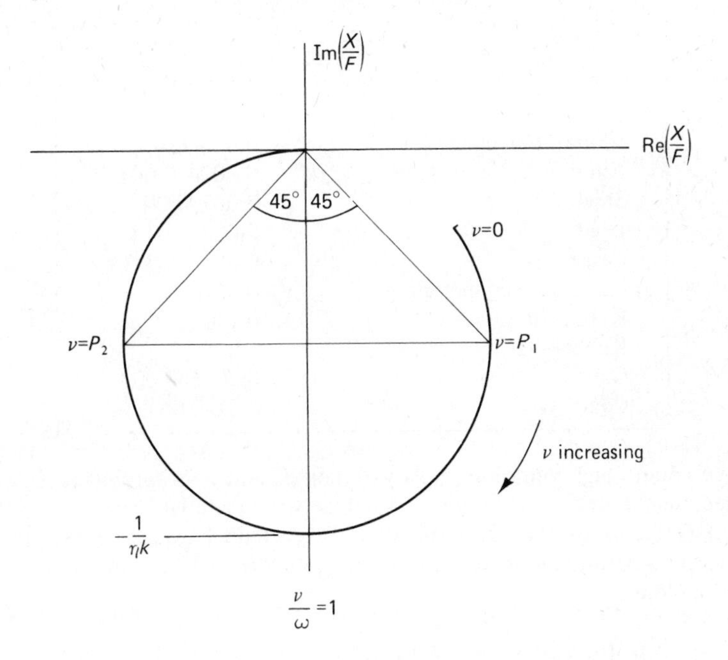

Fig. 5.5 – Receptance vector locus for system with hysteretic damping.

5.4 SOURCES OF DAMPING

The damping which occurs in structures can be considered to be either inherent damping, that is damping which occurs naturally within the structure or its environment, or added damping which is that resulting from specially constructed dampers added to the structure.

5.4.1 Inherent damping

5.4.1.1 Hysteretic or material damping

The load-extension hysteresis loops for linear materials and structures are elliptical under sinusoidal loading, and increase in area according to the square of the extension. Although the loss factor of a material depends on its composition, temperature, stress and the type of loading mechanism used, an approximate value for η can be obtained.

A range of values of η for some common engineering materials is given in Table 5.3. For more detailed information on material damping mechanisms and loss factors, see *Damping of Materials and Members in Structural Mechanisms* by B. J. Lazan (Pergamon).

High damping metals and alloys are often unsuitable for many engineering structures because of their low strength, ductility and hardness, and their high cost. Manganese copper is an exception in that it has a high ultimate strength, hardness and ductility. However, these special alloys are difficult to produce, and their damping is only large at high strains which means that structures would

Table 5.3

Material	Loss factor
Aluminium–pure	0.000 02–0.002
Aluminium alloy–dural	0.0004–0.001
Steel	0.001–0.008
Lead	0.008–0.014
Cast iron	0.003–0.03
Manganese copper alloy	0.05–0.1
Rubber–natural	0.1–0.3
Rubber–hard	1.0
Glass	0.0006–0.002
Concrete	0.01–0.06

have to endure high vibration levels with consequent noise problems. It will be realised that a steel or aluminium structure with material damping alone will have a Q factor of the order of 1000. This would be quite unacceptable in practice and fortunately rarely arises because additional damping occurs in the structural joints.

5.4.1.2 Damping in structural joints

The Q factor of a bolted steel structure is usually between 20 and 60, and for a welded steel structure a Q factor between 30 and 100 is common. Reinforced concrete can have Q factors in the range 15 to 25. Since the damping in the structural material is very small, most of the damping which occurs in real structures arises in the structural joints. However, even though over 90% of the inherent damping in most structures arises in the structural joints, little effort is made to optimise or even control this source of damping. This is because the energy dissipation mechanism in a joint is a complex process which is largely influenced by the interface pressure. At low joint clamping pressures sliding on a macro scale takes place and Coulomb's Law of Friction is assumed to hold. If the joint clamping pressure increases, mutual embedding of the surfaces starts to occur. Sliding on a macro-scale is reduced and micro-slip is initiated which involves very small displacements of an asperity relative to its opposite surface. A further increase in the joint clamping pressure will cause greater penetration of the asperities. The pressure on the contact areas will be the yield pressure of the softer material. Relative motion causes further plastic deformation of the asperities.

In most joints all three mechanisms operate, their relative significance depending on the joint conditions. In joints with high normal interface pressures and relatively rough surfaces, the plastic deformation is significant. Many joints have to carry pressures of this magnitude to satisfy criteria such as high static stiffness. A low normal interface pressure would tend to increase the significance of the slip mechanisms, as would an improvement in the quality of the surfaces in contact. With the macro-slip mechanism, the energy dissipation is proportional to the product of the interface shear force and the relative tangential motion. Under high pressure, the slip is small, and under low pressure the shear force is small: between these two extremes, the product becomes a maximum.

However, when two surfaces nominally at rest with respect to each other are subjected to slight vibrational slip, fretting corrosion can be instigated. This is a particularly serious form of wear inseparable from energy dissipation by interfacial slip, and hence frictional damping.

The fear of fretting corrosion occurring in a structural joint is one of the main reasons why joints are tightly fastened. However joint surface preparations such as cyanide hardening and electro-discharge machining are available which reduce the fretting corrosion from frictional damping in joints considerably, whilst allowing high joint damping. Plastic layers and greases have been used to separate the interfaces in joints and prevent fretting, but they have been squeezed out and have not been durable. Careful joint design and location is necessary if joint damping is to be increased in a structure without fretting corrosion becoming a problem; full details of fretting are given in *Fretting Fatigue* by R. B. Waterhouse (Applied Science Publishers).

The theoretical assessment of the damping which may occur in joints is difficult to make because of the variations in μ which occur in practice. However it is generally accepted that the friction force generated between the joint interfaces is usually:

(i) Dependent on the materials in contact and their surface preparation,
(ii) Proportional to the normal force across the interface,
(iii) Substantially independent of the sliding speed and apparent area of contact, and
(iv) Greater just prior to the occurrence of relative motion than during uniform relative motion.

The equations of motion of a structure with friction damping are thus non-linear: most attempts at analysis linearise the equations in some way. A very useful method is to calculate an equivalent viscous damping coefficient such that the energy dissipated by the friction and viscous dampers is the same. This has been shown to give an acceptable qualitative analysis for macro-slip. Some improvement on this method can be obtained by replacing μ by a term which allows for changes in the coefficient with slip amplitude. Some success has also been obtained by simply replacing the friction force with an equivalent harmonic force which is, essentially, the first term of the Fourier series representing the periodic friction force.

Some effects of controlling the joint clamping forces in a structure can be seen by considering an elastically supported beam fitted with friction joints at each end, as shown in Fig. 5.6.

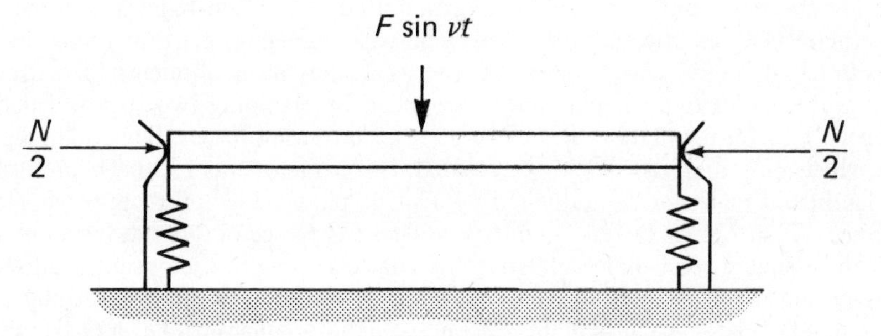

Fig. 5.6 – Elastically supported beam with coulomb damping.

The beam is excited by the harmonic force $F \sin \nu t$ applied at mid span. When the friction joints are very slack, $N \triangleq 0$ and the beam responds as an elastic beam on spring supports. When the joints are very tight, $N \triangleq \infty$ and the beam responds to excitation as if built-in at each end. For $\infty > N > 0$, a damped response occurs such as that shown in Fig. 5.7.

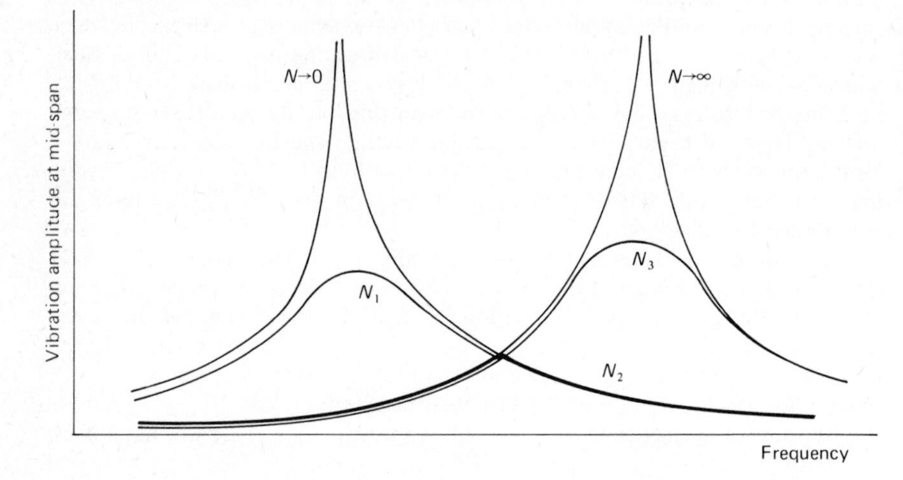

Fig. 5.7 – Amplitude-frequency response for beam shown in Fig. 5.6.

If N is increased from zero to N_1, a damped elastic response is achieved; significant damping occurs only when the beam vibration is sufficient to cause relative slip in the joints. As N approaches N_2, the beam responds as if built-in, until a vibration amplitude is reached when the joints slip, and the response is the same as that for the damped-elastic beam. When N increases to N_3, the built-in beam response is maintained until a higher amplitude is reached before slip takes place. The minimum response is achieved when $N = N_2$. This is obviously a powerful technique for controlling the dynamic response of structures, since both the maximum response, and the frequency at which this occurs, can be optimised. Friction damping can also be applied to joints that slip in rotation as well as, or instead of, translation.

The damping in plate type structures and elements can be increased by fabricating the plate out of several laminates bolted or rivetted together, so that as the plate vibrates interfacial slip occurs between the laminates thus giving rise to frictional damping. A Q factor as low as 20 has been obtained for a freely supported laminated circular plate, produced by clamping two identical plates together to form a plate subjected to interfacial friction forces. For a solid plate, in which only material damping occurred, the Q factor was 1300. Theoretically a laminated plate can be modelled by a single plate subjected to in-plane shear forces. When tightly fastened along two edges a Q factor of 345 has been obtained for a square steel plate; adjusting the edge clamping to the optimum allowed the Q factor to fall to 15, for the first mode of vibration. Replacing the plate by two similar plates, each half the thickness of the original enabled a Q factor of 75 to be achieved, even when the edges were tightly clamped. This improved to a

Q of 25 when optimum edge clamping was applied. However, some loss in stiffness must be expected, leading to a reduction in the resonance frequencies.

This technique has been applied with some success to plate type structural elements such as engine oil sumps, for reducing the noise and vibration generated.

It is often unnecessary to add a special damping device to a structure to increase the frictional damping, optimisation of an existing joint or joints being all that is required. Thus it can be cheap and easy to increase the inherent damping in a structure by optimising the damping in joints, although careful design is sometimes necessary to ensure that an adequate stiffness is maintained. It must be recognised that for joint damping to be large, slip must occur, and that fretting corrosion and joint damping are inseparable. Furthermore, some of the stiffness of a tightly clamped structure must be sacrificed if this source of damping is to be increased, although this loss in stiffness need not be large if the joints are carefully selected. This damping mechanism is most effective at low frequencies and the first few modes of vibration, since only under these conditions are the vibration amplitudes generally large enough to allow significant slip, and therefore damping, from this mechanism.

5.4.1.3 Other damping sources

In general the major sources of damping in a structure are that occurring in the joints and the structural material. Occasionally however structures are required to work in environments which contribute significantly to the total damping. For example ship hulls benefit from considerable hydrodynamic damping from the water, and this is true for all water immersed structures, and aerodynamic damping, though itself small, may be important in lightly damped structures.

5.4.2 Added damping

When the inherent damping in a structure is insufficient, it can be increased either by adding vibration dampers to the structure or by manufacturing the structure, or a part of it, out of a layered material with very high damping properties.

5.4.2.1 Vibration dampers and absorbers

A wide range of damping devices are commercially available; these may rely on viscous, dry friction or hysteretic effects. In most cases some degree of adjustment is provided, although the effect of the damper can usually be fairly well predicted by using the above theory. The viscous type damper is usually a cylinder with a closely fitting piston and filled with a fluid. Suitable valves and porting give the required resistance to motion of the piston in the cylinder. Dry friction dampers rely on the friction force generated between two or more surfaces pressed together under a controlled force. Hysteretic type dampers are usually made from an elastic material with high internal damping, such as natural rubber. Occasionally dampers relying on other effects such as eddy currents are used.

However, these added dampers only act to reduce the vibration of a structure. If a particularly troublesome resonance exists it may be preferable to add a **vibration absorber**. This is simply a spring-body system which is added to the structure; the parameters of the absorber are chosen so that the amplitude of the vibration of the structure is greatly reduced, or even eliminated, at a frequency which is usually chosen to be at the original troublesome resonance. Consider the model of the structure shown in Fig. 5.8, where K and M are the effective stiffness and mass of the structure when vibrating in the troublesome mode.

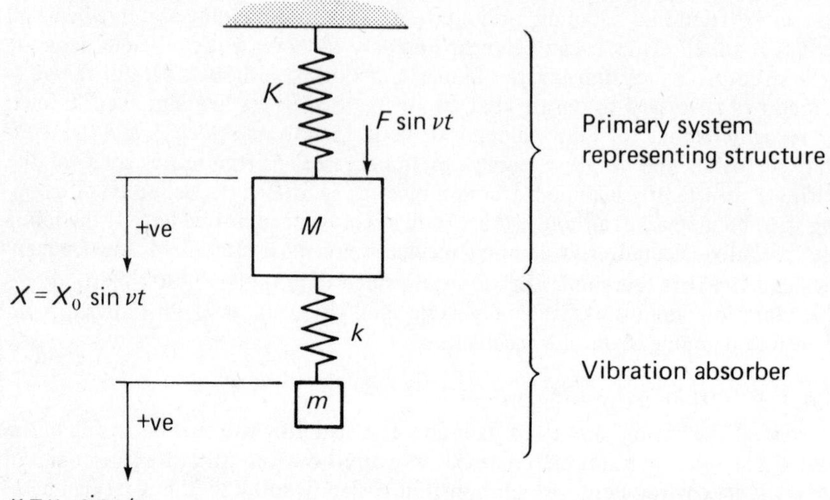

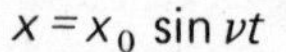

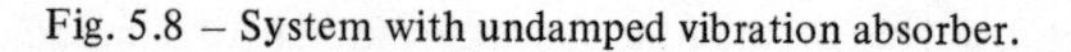

Fig. 5.8 – System with undamped vibration absorber.

The absorber is represented by the system with parameters k and m. From section 3.1.3 it can be seen that the equations of motion are

$$M\ddot{X} = -KX - k(X - x) + F \sin \nu t,$$

and $$m\ddot{x} = k(X - x).$$

Substituting $X = X_0 \sin \nu t$ and $x = x_0 \sin \nu t$,

$$X_0(K + k - M\nu^2) + x_0(-k) = F,$$

and $$X_0(-k) + x_0(k - m\nu^2) = 0.$$

Thus $$X_0 = \frac{F(k - m\nu^2)}{\Delta},$$

and $$x_0 = \frac{Fk}{\Delta},$$

where $$\Delta = (k - m\nu^2)(K + k - M\nu^2) - k^2,$$

and $\Delta = 0$ is the frequency equation.

It can be seen that not only does the system now possess two natural frequencies Ω_1 and Ω_2 instead of one, but by arranging for $k - m\nu^2 = 0$, X_0 can be made zero.

Thus if $\sqrt{k/m} = \sqrt{K/M}$, the response of the primary system at its original resonance frequency can be made zero. This is the usual tuning arrangement for an undamped absorber because the resonance problem in the primary system is only severe when $\nu \simeq \sqrt{K/M}$ rad/s. This is shown in Fig. 5.9.

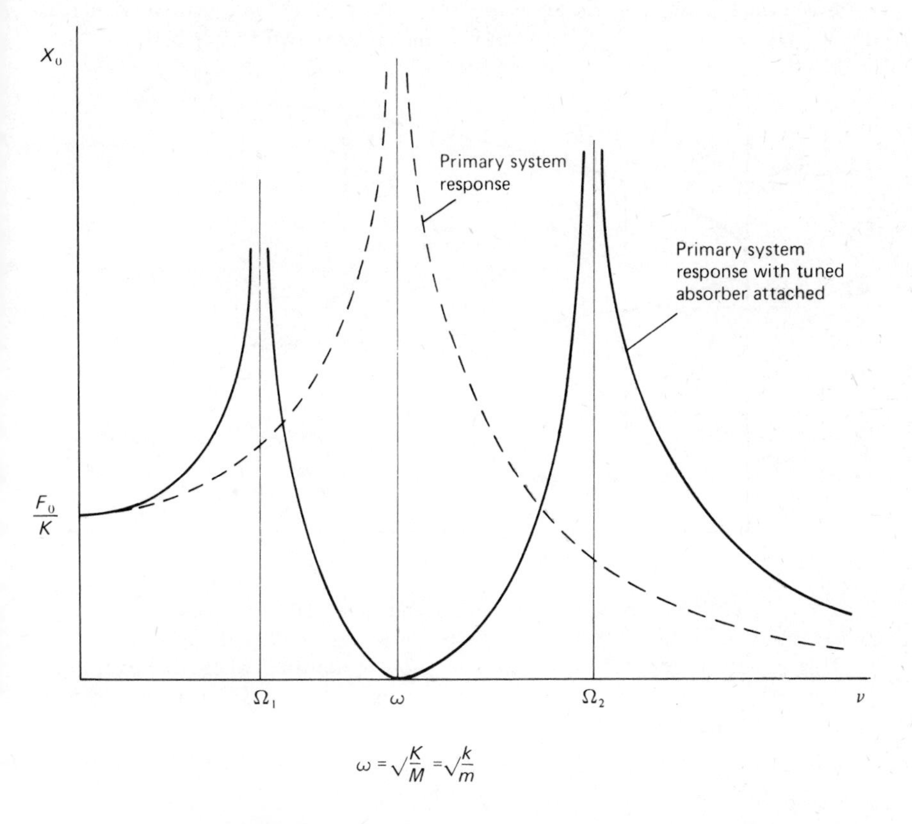

Fig. 5.9 – Amplitude-frequency response for system with and without tuned absorber.

When $X_0 = 0$, $x_0 = -F/k$, so that the force in the absorber spring, kx_0 is $-F$, thus the absorber applies a force to the primary system which is equal and opposite to the exciting force. Hence the body in the primary system has a net zero exciting force acting on it and therefore zero vibration amplitude.

If an absorber is correctly tuned $\omega^2 = K/M = k/m$, and if the mass ratio $\mu = m/M$, the frequency equation $\Delta = 0$ is:

$$\left(\frac{\nu}{\omega}\right)^4 - (2+\mu)\left(\frac{\nu}{\omega}\right)^2 + 1 = 0.$$

Hence
$$\left(\frac{\nu}{\omega}\right)^2 = \left(1 + \frac{\mu}{2}\right) \pm \sqrt{\mu + \frac{\mu^2}{4}}$$

and the natural frequencies Ω_1 and Ω_2 are found to be

$$\frac{\Omega_{1,2}}{\omega} = \left[\left(1 + \frac{\mu}{2}\right) \pm \sqrt{\mu + \frac{\mu^2}{4}}\right]^{1/2}.$$

For a small μ, Ω_1 and Ω_2 are very close to each other, and near to ω; increasing μ gives better separation between Ω_1 and Ω_2 as shown in Fig. 5.10.

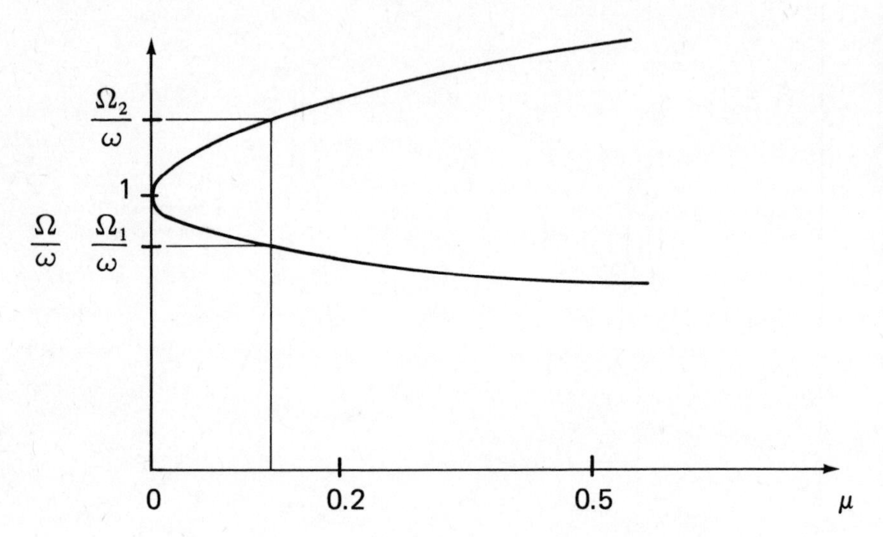

Fig. 5.10 – Effect of absorber mass ratio on natural frequencies.

This effect is of great importance in those structures where the excitation frequency may vary; if μ is small resonances at Ω_1 or Ω_2 may be excited. It should be noted that since

$$\left(\frac{\Omega_1}{\omega}\right)^2 = \left(1 + \frac{\mu}{2}\right) - \sqrt{\mu + \frac{\mu^2}{4}},$$

and $$\left(\frac{\Omega_2}{\omega}\right)^2 = \left(1 + \frac{\mu}{2}\right) + \sqrt{\mu + \frac{\mu^2}{4}},$$

Then $$\frac{\Omega_1{}^2\Omega_2{}^2}{\omega^4} = \left(1 + \frac{\mu}{2}\right)^2 - \left(\mu + \frac{\mu^2}{4}\right) = 1$$

That is $\Omega_1 . \Omega_2 = \omega^2$.

Also $$\left(\frac{\Omega_1}{\omega}\right)^2 + \left(\frac{\Omega_2}{\omega}\right)^2 = 2 + \mu.$$

Example 20

A system has a violent resonance at 79 Hz. As a trial remedy a vibration absorber is attached which results in a resonance frequency of 65 Hz. How many such absorbers are required if no resonance is to occur between 60 and 120 Hz?

Since $$\left(\frac{\Omega_1}{\omega}\right)^2 + \left(\frac{\Omega_2}{\omega}\right)^2 = 2 + \mu,$$

and $$\Omega_1 . \Omega_2 = \omega^2$$

in the case of one absorber, with $\omega = 79$ Hz and $\Omega_1 = 65$ Hz,

$$\Omega_2 = \frac{79^2}{65} = 96 \text{ Hz.}$$

Also $$\left(\frac{65}{79}\right)^2 + \left(\frac{96}{79}\right)^2 = 2 + \mu,$$

Thus $\mu = 0.154.$

In case of n absorbers,

if $$\Omega_1 = 60 \text{ Hz}, \Omega_2 = \frac{79^2}{60} = 104 \text{ Hz (too low)}$$

So require $\Omega_2 = 120$ Hz and $\Omega_1 = \dfrac{79^2}{120} = 52$ Hz.

Hence $$\left(\frac{52}{79}\right)^2 + \left(\frac{120}{79}\right)^2 = 2 + \mu.$$

Thus $$\mu' = 0.74$$
$$= n.\mu,$$

and $$n = \frac{0.74}{0.154} = 4.82.$$

Thus 5 absorbers are required.

Example 21

A machine tool of mass 3000 kg has a large resonance vibration in the vertical direction at 120 Hz. To control this resonance, an undamped vibration absorber of mass 600 kg is fitted, tuned to 120 Hz. Find the frequency range in which the amplitude of the machine vibration is less with the absorber fitted than without.

If (X_0) with absorber = (X_0) without absorber,

$$\frac{F(k - mv^2)}{(K + k - Mv^2)(k - mv^2) - k^2} = -\frac{F}{K - Mv^2}$$

Multiplying out and putting $\mu = m/M$ gives

$$2\left(\frac{v}{\omega}\right)^4 - (4 + \mu)\left(\frac{v}{\omega}\right)^2 + 2 = 0.$$

Since $\mu = \dfrac{600}{3000} = 0.2,$

$$\left(\frac{\nu}{\omega}\right)^2 = \frac{4+\mu}{4} \pm \tfrac{1}{4}\sqrt{\mu^2 + 8\mu} = 1.05 \pm 0.32.$$

Thus $\dfrac{\nu}{\omega} = 1.17$ or $0.855,$

and $\nu = 102$ Hz or 140 Hz.

Thus the required frequency range is 102–140 Hz.

A convenient analysis of a system with a vibration absorber can be carried out by using the receptance technique.

Consider the undamped dynamic vibration absorber shown in Fig. 5.11. The system is split into sub-systems A and B, where B represents the absorber.

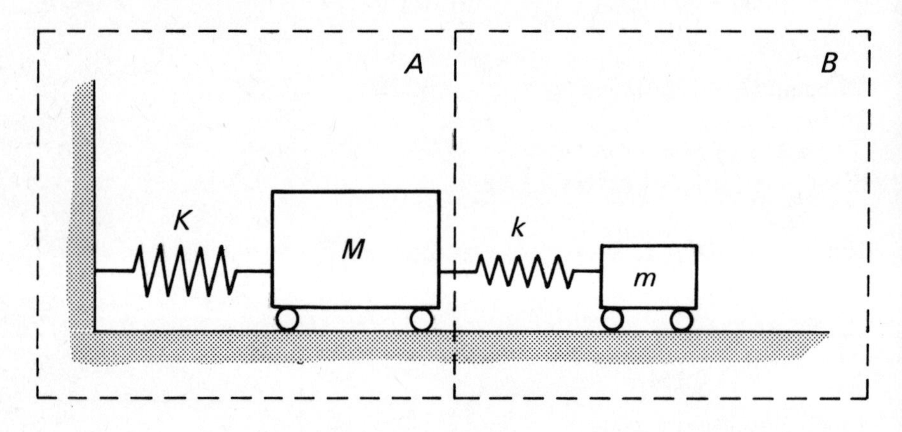

Fig. 5.11 – Sub-system analysis.

For sub-system A, (the structure),

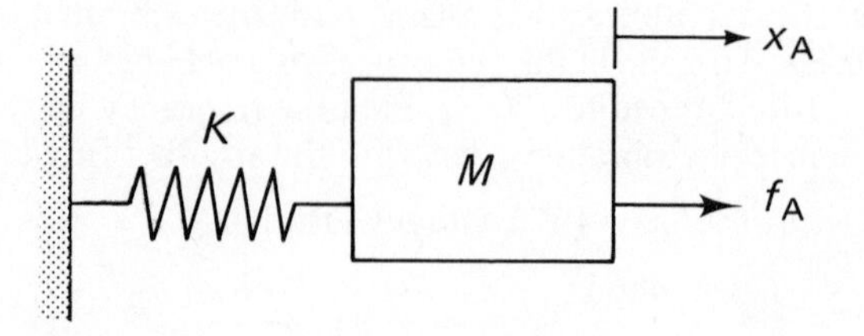

$$f_A = M\ddot{x}_A + Kx_A,$$

and $$\alpha = \frac{1}{K - M\nu^2}.$$

For sub-system B, (the absorber),

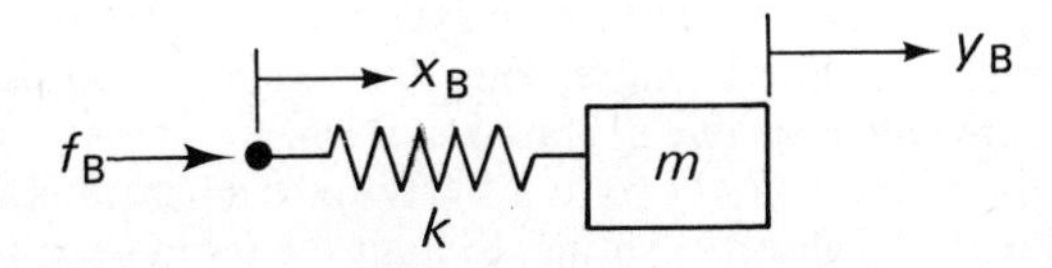

$$f_B = k(x_B - y_B) = m\ddot{y}_B = -m\nu^2 Y_B,$$

and $$\beta = -\left(\frac{k - m\nu^2}{km\nu^2}\right)$$

Thus the frequency equation $\alpha + \beta = 0$ gives

$$Mm\nu^4 - (mK + Mk + mk)\,\nu^2 + Kk = 0, \text{ as before.}$$

It is often convenient to solve the frequency equation $\alpha + \beta = 0$ or $\alpha = -\beta$ by a graphical method. In the case of the absorber, both α and $-\beta$ can be plotted as a function of ν, and the intersections give the natural frequencies Ω_1 and Ω_2, as shown in Fig. 5.12.

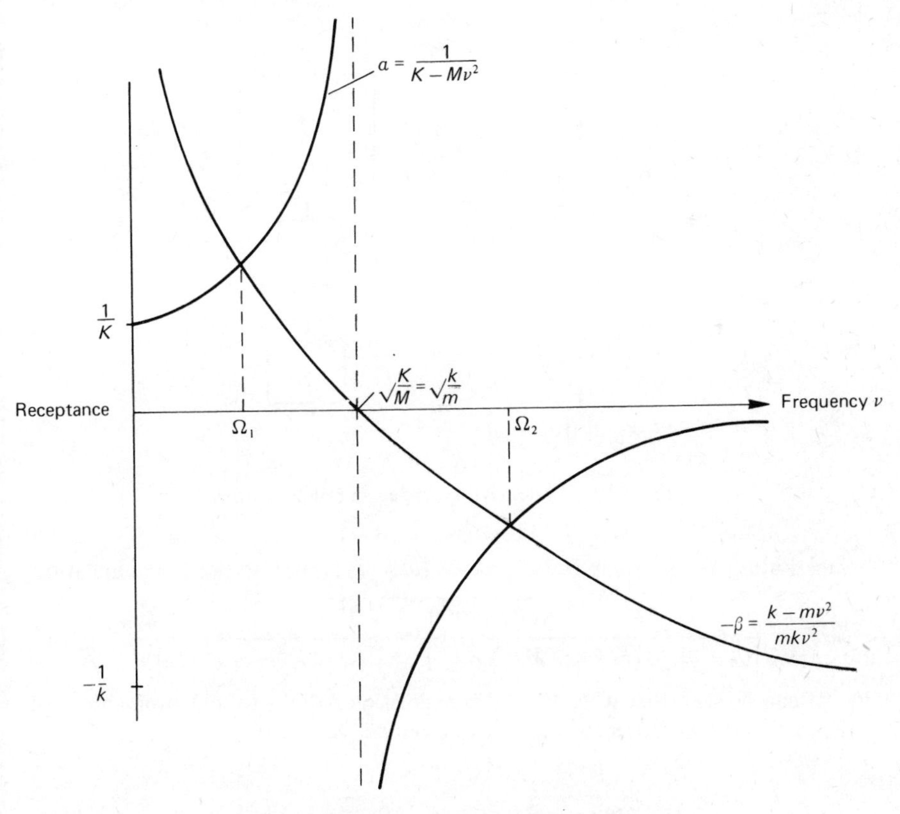

Fig. 5.12 – Sub-system receptance-frequency responses.

This technique is particularly useful when it is required to investigate the effect of several different absorbers, since once the receptance of the primary system is known, it is only necessary to analyse each absorber and not the complete system in each case. Furthermore, sometimes the receptances of structures are measured and are only available in graphical form.

If the proximity of Ω_1 and Ω_2 to ω is likely to be a hazard, damping can be added in parallel with the absorber spring, to limit the response at these frequencies. Unfortunately, if damping is added, the response at the frequency ω will no longer be zero.

Fig. 5.13 shows the primary system with a viscous damped absorber added. The equations of motion are

$$M\ddot{X} = F \sin \nu t - KX - k(X - x) - c(\dot{X} - \dot{x}),$$

and

$$m\ddot{x} = k(X - x) + c(\dot{X} - \dot{x}).$$

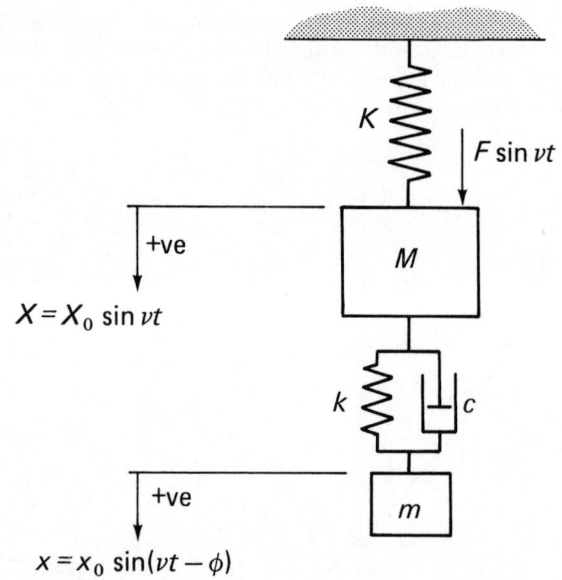

Fig. 5.13 – System with damped vibration absorber.

Substituting $X = X_0 \sin \nu t$ and $x = x_0 \sin(\nu t - \phi)$ gives, after some manipulation,

$$X_0 = \frac{F\sqrt{(k - m\nu^2)^2 + (c\nu)^2}}{\sqrt{[(k - m\nu^2)(K + k - M\nu^2) - k^2]^2 + [(K - M\nu^2 - m\nu^2)c\nu]^2}}.$$

It can be seen that when $c = 0$ this expression reduces to that given above for the undamped vibration absorber. Also when c is very large

$$X_0 = \frac{F}{K - (M + m)\nu^2}$$

For intermediate values of c the primary system response has damped resonance peaks, although the amplitude of vibration does not fall to zero at the original resonance frequency. This is shown in Fig. 5.14.

The response of the primary system can be minimized over a wide range of exciting frequencies by carefully choosing the value of c, and also arranging the system parameters so that the points P_1 and P_2 are at about the same amplitude. However one of the main advantages of the undamped absorber, that of reducing the vibration amplitude of the primary system to zero at the troublesome resonance, is lost.

A design criteria that has to be carefully considered is the possible fatigue and failure of the absorber spring: this could have severe consequences. In view of this, some damped absorber systems dispense with the absorber spring and sacrifice some of the absorber effectiveness. This has particularly wide application in torsional systems, where the device is known as a **Lanchester Damper**.

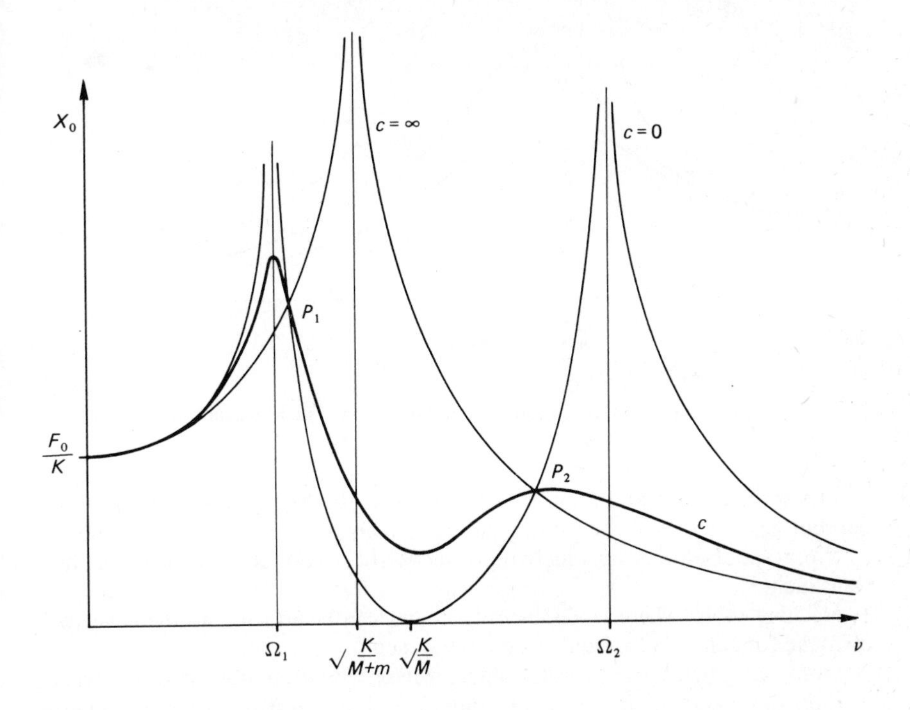

Fig. 5.14 – Effect of absorber damping on system response.

It can be seen that if $k = 0$,

$$X_0 = \frac{F\sqrt{m^2\nu^4 + c^2\nu^2}}{\sqrt{[(K - M\nu^2)\,m\nu^2]^2 + [(K - M\nu^2 - m\nu^2)\,c\nu]^2}} .$$

When $c = 0$, $X_0 = \dfrac{F}{K - M\nu^2}$ (no absorber)

and when c is very large, $X_0 = \dfrac{F}{K - (M + m)\,\nu^2}$.

These responses are shown in Fig. 5.15 together with that for the optimum value of c.

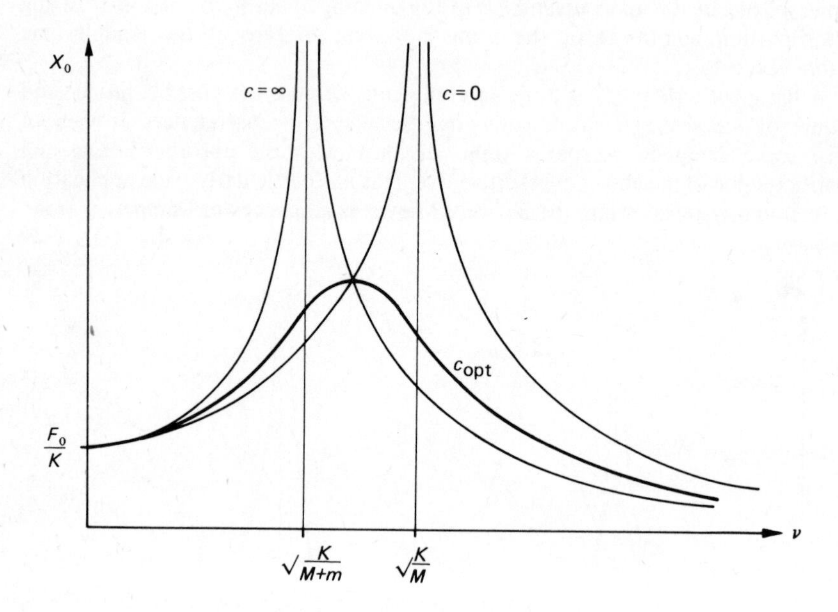

Fig. 5.15 – Effect of Lanchester damper on system response.

The springless vibration absorber is much less effective than the sprung absorber, but has to be used when spring failure is likely, or would prove disastrous.

Vibration absorbers are widely used to control structural resonances. Applications include,

1. Machine tools, where large absorber bodies can be attached to the headstock or frame for control of a troublesome resonance.
2. Overhead power transmission lines, where vibration absorbers known as Stockbridge dampers are used for controlling line resonance excited by cross winds.
3. Engine crankshaft torsional vibration, where Lanchester dampers can be attached to the pulley for the control of engine harmonics.
4. Footbridge structures, where pedestrian excited vibration has been reduced by an order of magnitude by fitting vibration absorbers.
5. Engines, pumps and diesel generator sets where vibration absorbers are fitted so that the vibration transmitted to the supporting structure is reduced or eliminated.

Not all damped absorbers rely on viscous damping; dry friction damping is often used, and the replacement of the spring and damper elements by a single rubber block possessing both properties is fairly common.

A structure or mechanism which has loosely fitting parts is often found to rattle when vibration takes place. Rattling consists of a succession of impacts, these dissipate vibrational energy and therefore rattling increases the structural damping. It is not desirable to have loosely fitting parts in a structure, but an impact damper can be fitted.

An **impact damper** is a hollow container with a loosely fitting body or slug; vibration causes the slug to impact on the container ends thereby dissipating vibrational energy. The principle of the impact damper is that when two bodies collide some of their energy is converted into heat and sound so that the vibrational energy is reduced. Sometimes the slug is supported by a spring so that advantage can be taken of resonance effects. Careful tuning is required, particularly with regard to slug mass, material and clearance, if the optimum effect is to be achieved. Although cheap and easy to manufacture and install, impact dampers have been neglected because they are difficult to analyse and design, and their performance can be unpredictable. They are also rather noisy in operation, although the use of PVC impact surfaces can go some way to reducing this. Some success has been achieved by fitting vibration absorbers with impact dampers. The significant advantage of the impact vibration absorber over the conventional dynamic absorber is the reduction of the amplitude of the primary system both at resonance and at higher frequencies.

5.4.2.2 Constrained layer damping

The polymer materials which exist that have very high damping properties lack sufficient rigidity and creep resistance to enable a structure to be fabricated from them, so that if advantage is to be taken of their high damping a composite construction of a rigid material, such as a metal, with damping layers bonded to it has to be used, usually as a beam or plate. High damping material can be applied to a structure by fabricating it, at least in part, from elements in which layers of high damping viscoelastic material are bonded between layers of metal. When the composite material vibrates the constrained damping layers are subjected to shear effects, which cause vibrational energy to be converted into heat and hence dissipated. Other applications of high damping polymers are to edge damping, where the polymer forms the connection between a panel or beam and its support, and unconstrained layers, where the damping material is simply bonded to the surface of the structural element. Whilst these applications do increase the total damping, they are not as effective as using constrained layers.

Before considering the damping effects that can be achieved by the constrained layer technique, it must be emphasised that the properties of viscoelastic materials are both temperature and frequency sensitive. Fig. 5.16 shows how the shear modulus and loss factor can vary.

Another disadvantage with composite materials is that they are difficult to bend or form without reducing their damping capabilities, because of the distortion that occurs in the damping layer.

Two, three, four or five or more layers of viscoelastic material and metal can be used in a composite; each layer can have particular properties, thickness and location relative to the neutral axis so that the composite as a whole has the most desirable structural and dynamic performance. Because of this wide variation in composite material geometry, only a three layer symmetrical construction will be considered, other geometries being an extension of the three layer composite.

Consider the composite beam of length l shown in Fig. 5.17.

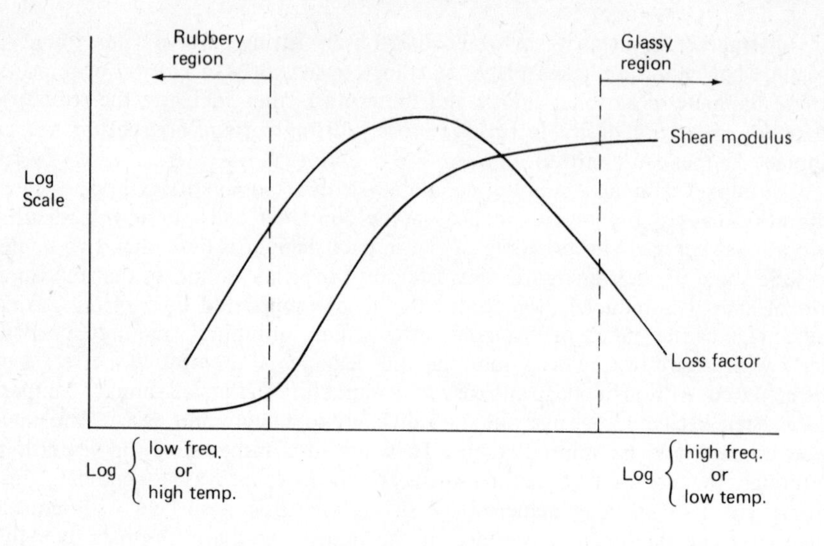

Fig. 5.16 – Viscoelastic material properties.

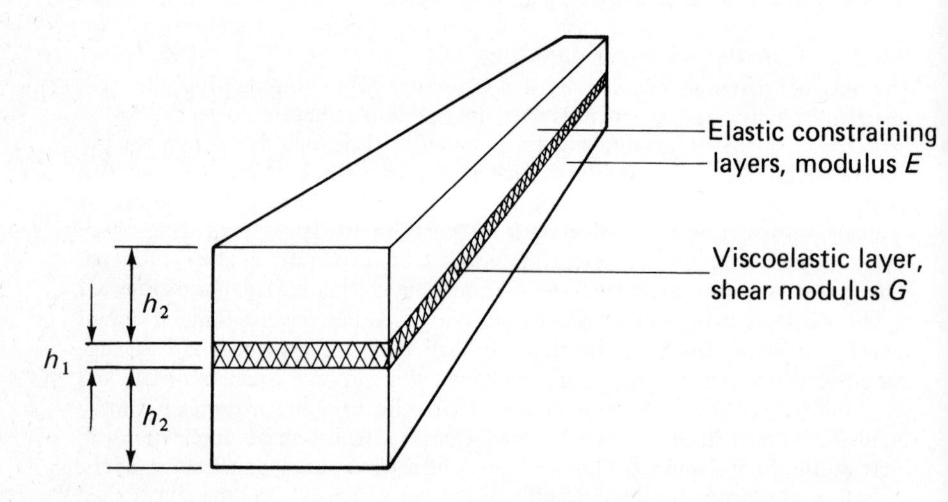

Fig. 5.17 – Composite beam.

A dimensionless shear parameter can be defined, equal to

$$\frac{2l^2G}{h_1h_2E}.$$

It is assumed that the elastic constraining layers have a zero loss factor, and that the viscoelastic damping layer has zero stiffness. The beam loss factor for a cantilever vibrating in its first mode is shown as a function of the shear parameter in Fig. 5.18, for various values of the loss factor η for the viscoelastic layer material.

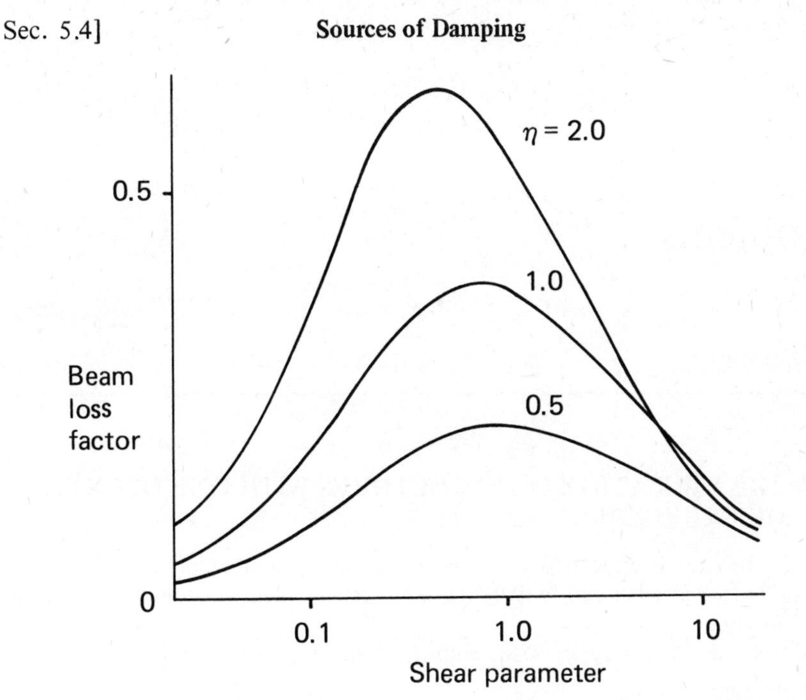

Fig. 5.18 – Effect of layer loss factor on beam loss factor as a function of the shear parameter

It can be seen from Fig. 5.18 that a high beam loss factor is only obtainable at a particular value of the shear parameter, and that as the loss factor of the viscoelastic layer increases the curves become sharper. The dependance of the beam loss factor on the shear parameter is consequently of great practical significance. However very high beam loss factors can be obtained resulting in a Q value of two or even less.

The difference between the optimum loss factors for the first three modes of a cantilever has been shown to be less than 10%. Most viscoelastic materials have a shear modulus which increases with frequency, so that the damping can be kept near to the optimum over a large frequency range. It must be emphasised that it is not possible to secure high structural damping and high stiffness by this method of damping.

Damping in structures, and constrained layer damping in particular, has been discussed in *Structural Damping* by J. E. Ruzicka (Pergamon Press), and in *Damping Applications for Vibration Control* edited by P. J. Torvik (ASME Publication AMD, vol. 38).

Chapter 6

Problems

6.1 THE VIBRATION OF STRUCTURES WITH ONE DEGREE OF FREEDOM

1. A structure is modelled by a rigid horizontal member of mass 3000 kg, supported at each end by a light elastic vertical member of flexural stiffness 2 MN/m.
   Find the frequency of small amplitude horizontal vibrations of the rigid member.
2. Part of a structure is modelled by a thin rigid rod of mass m pivoted at the lower end, and held in the vertical position by two springs, each of stiffness k, as shown.
   Find the frequency of small amplitude oscillation of the rod about the pivot.

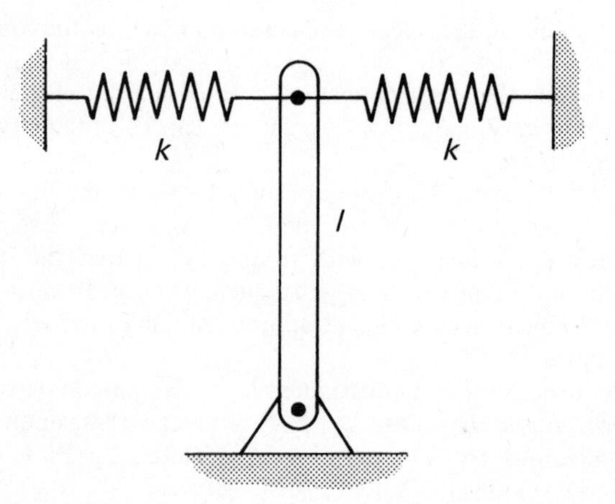

3. A unform beam of length 8 m, simply supported at the ends, carries a uniformly distributed mass of 300 kg/m and three bodies, one of mass 1000 kg at mid-span and two of mass 1500 kg each, at 2 m from each end. The second moment of area of the beam is 10^{-4} m^4 and the modulus of elasticity of the material is 200 GN/m^2.

Estimate the lowest natural frequency of flexural vibration of the beam assuming that the deflection y_x at a distance x from one end is given by:

$$y_x = y_c . \sin \pi (x/l),$$

where y_c is the deflection at mid-span and l is the length of the beam.

4. A section of steel pipe in a distillation plant is 120 mm diameter, 10 mm thick and 5 m long. The pipe may be assumed to be built in at each end, so that the deflection y, at a distance x from one end of a pipe of length l, is

$$y = \frac{mg}{24\, EI} . x^2 (l - x)^2 ,$$

m being the mass per unit length.
Calculate the lowest natural frequency of transverse vibration of the pipe when full of water. Take $\rho_{steel} = 7750\ kg/m^3$, $\rho_{water} = 930\ kg/m^3$ and $E_{steel} = 200\ GN/m^2$.

5. A uniform horizontal steel beam is built in to a rigid structure at one end and pinned at the other end; the pinned end cannot move vertically but is otherwise unconstrained. The beam is 8 m long, the relevant flexural second moment of area of the cross-section is $4.3 \times 10^6\ mm^4$, and the beam's own mass together with the mass attached to the beam is equivalent to a uniformly distributed mass of 600 kg/m.
Using a combination of sinusoidal functions for the deflected shape of the beam, estimate the lowest natural frequency of flexural vibrations in the vertical plane.

6. Estimate the lowest frequency of natural transverse vibration of a chimney 100 m high, which can be represented by a series of lumped masses M at distances y from its base as follows:

y (m)	20	40	60	80	100
M (10^3 kg)	700	540	400	280	180

With the chimney considered as a cantilever on its side the state deflection in bending, x along the chimney is calculated to be

$$x = X \left(1 - \cos \pi \frac{y}{2l}\right),$$

where $l = 100$ m,

and $X = 0.2$ m.

How would you expect the actual frequency to compare with the frequency that you have calculated?

7. Estimate the lowest natural frequency of a light beam 7 m long carrying six concentrated masses equally spaced along its length. The measured static deflections under each mass are:

Mass (kg)	1070	970	370	370	670	670
Deflection (mm)	2.5	2.8	5.5	5.0	2.5	1.0

8. A uniform rigid building, height 30 m and cross section 10 m × 10 m, rests on an elastic soil of stiffness 0.6×10^6 N/m^3. (Stiffness defined as the force per unit area to produce unit deflection).
If the mass of the building is 2×10^6 kg and its inertia about its axis of rocking at the base is 500×10^6 kg m^2, calculate the period of the rocking motion (small amplitudes).
What wind speed would excite this motion if the Strouhal number is 0.22? Calculate also the maximum height the building could be before becoming unstable.
9. A single degree of freedom system with a body of mass 10 kg, a spring of stiffness 1 kN/m, and negligible damping, is subjected to an input force F which varies with time as shown below.

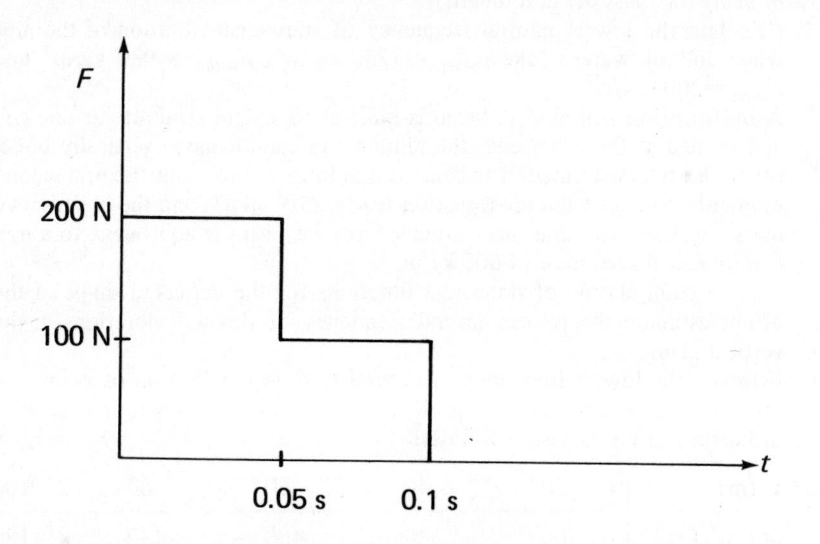

Determine the amplitude of free vibration of the body after the force is removed.
10. A uniform rigid tower of height 30 m and cross section 3 m × 3 m, is symmetrically mounted on a rigid foundation of depth 2 m and section 5 m × 5 m. The mass of the tower is calculated to be 1.5×10^5 kg, and of the foundation, 10^3 kg. The foundation rests on an elastic soil which has a uniform stiffness of 2×10^6 N/m^3. (Stiffness defined as the force per unit area to produce unit deflection.)
If the mass moment of inertia of the tower and foundation about its axis of rocking at the base of the foundation is 6×10^7 kg m^2, find the period of small amplitude rocking motion. The axis of rocking is parallel to a side of the foundation.
What is the greatest height the tower could have and still be stable on this foundation?
11. The foundation of a rigid tower is a circular concrete block of diameter D, set into an elastic soil. The effective stiffness of the soil, k, is defined as the force per unit area to produce a unit displacement and is constant for small deflections. The tower is uniform with a total mass M. The centre of mass is

situated on the centre line at a height h above the base. The moment of inertia of the tower about an axis of rocking at the base is I.
Show that the natural frequency of rocking is given by:

$$\frac{1}{2\pi}\sqrt{\frac{\pi k D^4/64 - Mgh}{I}} \text{ Hz.}$$

12. A body supported by an elastic structure performs a damped oscillation of period 1 s, in a medium whose resistance is proportional to the velocity. At a given instant the amplitude was observed to be 100 mm, and in 10 s this had reduced to 1 mm.
What would be the period of the free vibration if the resistance of the medium were negligible?
13. To determine the amount of damping in a bridge it was set into vibration in the fundamental mode by dropping a weight on it at centre span. The observed frequency was 1.5 Hz, and the amplitude was found to have decreased to 0.9 of the initial maximum after 2 seconds. The equivalent mass of the bridge (estimated by the Rayleigh Energy method) was 10^5 kg.
Assuming viscous damping and simple harmonic motion, calculate the damping coefficient, the logarithmic decrement and the damping ratio.
14. A new concert hall is to be protected from the ground vibrations from an adjacent highway by mounting the hall on rubber blocks. The predominant frequency of the sinusoidal ground vibrations is 40 Hz, and a motion transmissibility of 0.1 is to be attained at that frequency.
Calculate the static deflection required in the rubber blocks, assuming that these act as linear, undamped springs.
15. When considering the vibrations of a structure, what is meant by the Q factor? Derive a simple relationship between the Q factor and the damping ratio for a single degree of freedom system with light viscous damping.
Measurements of the vibration of a bridge section resulting from impact tests show that the period of each cycle is 0.6 s, and that the amplitude of the third cycle is twice the amplitude of the ninth cycle. Assuming the damping to be viscous estimate the Q factor of the section.
When a vehicle of mass 4000 kg is positioned at the centre of the section the period of each cycle increases to 0.62 s; no change is recorded in the rate of decay of the vibration. What is the effective mass of the section?
16. The vibration of the floor of a laboratory is found to be simple harmonic motion at a frequency in the range 15-60 Hz, (depending on the speed of some nearby reciprocating plant). It is desired to install in the laboatory a sensitive instrument which requires insulating from the floor vibration. The instrument is to be mounted on a small platform which is supported by three similar springs resting on the floor, arranged to carry equal loads; the motion is restrained to occur in a vertical direction only. The combined mass of the instrument and the platform is 40 kg: the mass of the springs can be neglected and the equivalent viscous damping ratio of the suspension is 0.2.
Calculate the maximum value for the spring stiffness, if the amplitude of the transmitted vibration is to be less than 10% of that of the floor vibration over the given frequency range.

17. A two-wheeled trailer of sprung mass 700 kg is towed at 60 km/h, along an undulating straight road whose surface may be considered sinusoidal. The distance from peak to peak of the road surface is 30 m, and the height from hollow to crest 0.1 m. The effective stiffness of the trailer suspension is 60 kN/m, and the shock absorbers, which provide linear viscous damping, are set to give a damping ratio of 0.67.
    Assuming that only vertical motion of the trailer is excited, find the absolute amplitude of this motion and its phase angle relative to the undulations.
18. Find the Fourier series representation of the following triangular wave:

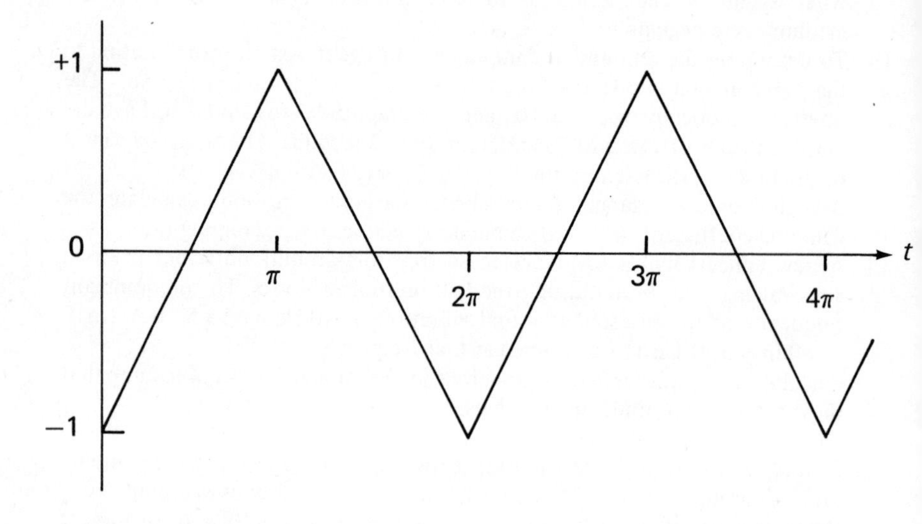

19. A wooden floor, 6 m by 3 m, is simply supported along the two shorter edges. The mass is 300 kg and the static deflection at the centre under the self-weight is 7 mm. It is proposed to determine the dynamic properties of the floor by dropping a sand bag of 50 kg mass on it at the centre, and to measure the response at that position with an accelerometer and a chart recorder.
    In order to select the instruments required, estimate:
    (i) the frequency of the fundamental mode of vibration which would be recorded,
    (ii) the number of oscillations at the fundamental frequency for the signal amplitude on the recorder to be reduced to half, assuming a loss factor of 0.05, and
    (iii) the height of drop of the sand bag so that the dynamic reflection shall not exceed 10 mm, and the corresponding maximum acceleration.

6.2 THE VIBRATION OF STRUCTURES WITH MORE THAN ONE DEGREE OF FREEDOM

20. A two-storey building is represented by the two degree of freedom lumped mass system shown below:

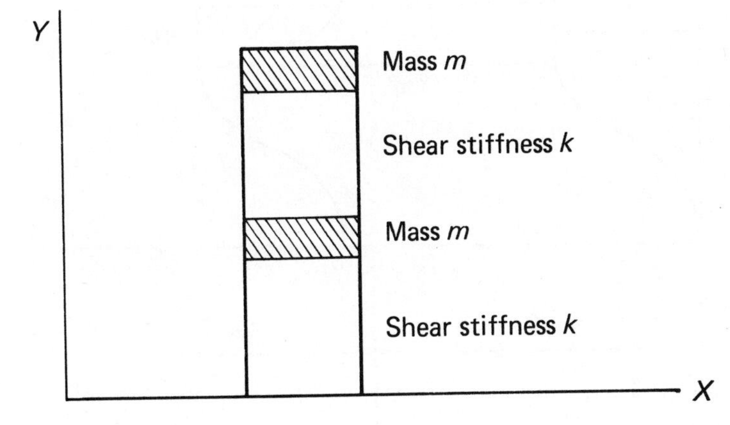

Obtain the frequency equation for swaying motion in the X–Y plane: hence calculate the natural frequencies and sketch the corresponding mode shapes.

21. A vehicle has a mass of 2000 kg and a 3 m wheelbase. The mass moment of inertia about the centre of mass is 500 kg m^2, and the centre of mass is located 1 m from the front axle. Considering the vehicle as a two degree of freedom system, find the natural frequencies and the corresponding modes of vibration, if the front and rear springs have stiffnesses of 50 kN/m and 80 kN/m respectively.
The expansion joints of a concrete road are 5 m apart. These joints cause a series of impulses at equal intervals to vehicles travelling at a constant speed. Determine the speeds at which pitching motion and up and down motion are most likely to arise for the above vehicle.

22. To analyse the vibrations of a two-storey building it is represented by the lumped mass system shown, where $m_1 = \frac{1}{2} m_2$, and $k_1 = \frac{1}{2} k_2$. (k_1 and k_2 represent the shear stiffnesses of the parts of the building shown.)
Calculate the natural frequencies of free vibrations, and sketch the corresponding mode shapes of the building, showing the amplitude ratios.
If a horizontal harmonic force $F_1 \sin \nu t$ is applied to the top floor, determine expressions for the amplitudes of the steady state vibration of each floor.

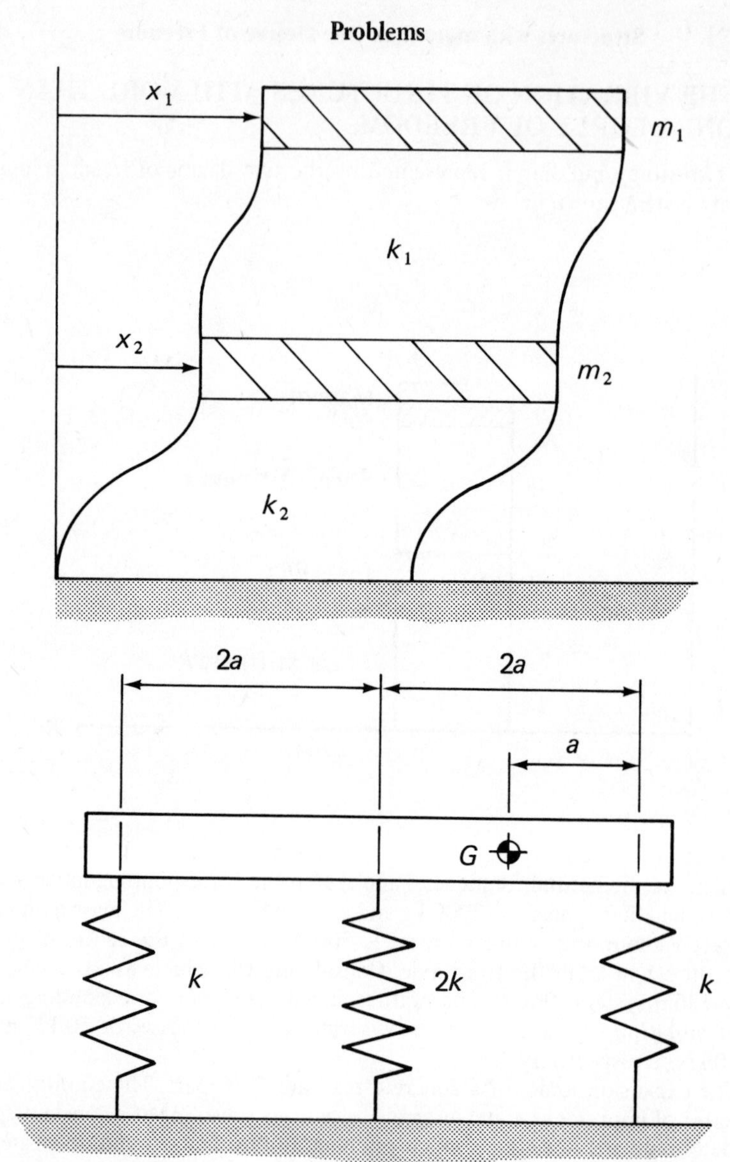

23.

2a 2a a G k 2k k

The rigid beam, shown in its position of static equilibrium in the figure, has a mass m and a mass moment of inertia $2\ ma^2$ about an axis perpendicular to the plane of the diagram, and through its centre of mass G.
Assuming no horizontal motion of G, find the frequencies of small oscillations in the plane of the diagram, and the corresponding positions of the nodes.

24. Part of a building structure is modelled by the triple pendulum shown. Obtain the equations of motion of small amplitude oscillation in the plane of the figure by using the Lagrange Equation.
Hence determine the natural frequencies of the structure and the corresponding mode shapes.

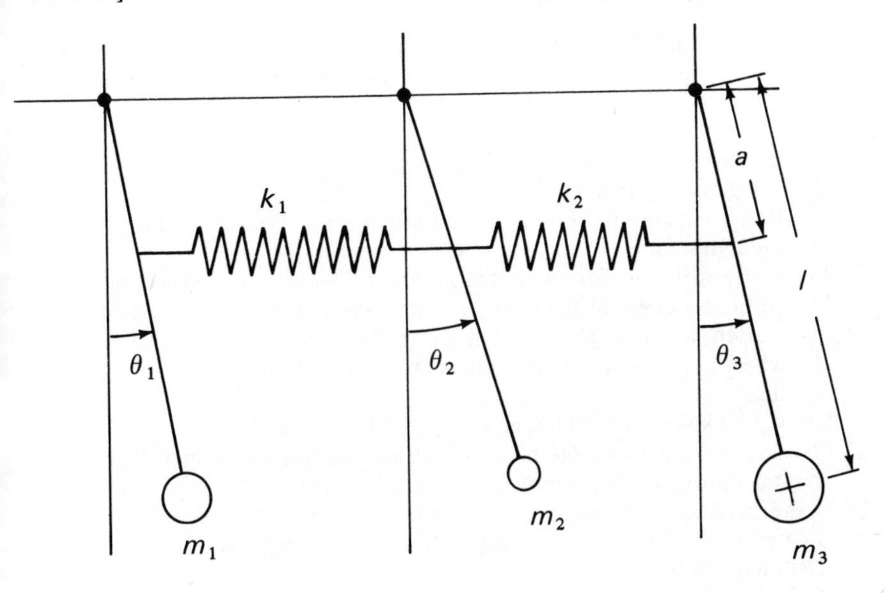

25. If the building in Problem 20 were enlarged by adding a further floor of mass m and shear stiffness k on top of the existing building, obtain the frequency equation for the three degree of freedom system formed. Given m and k contemplate how this equation may be solved. What if the building were 20 storeys high?

26.

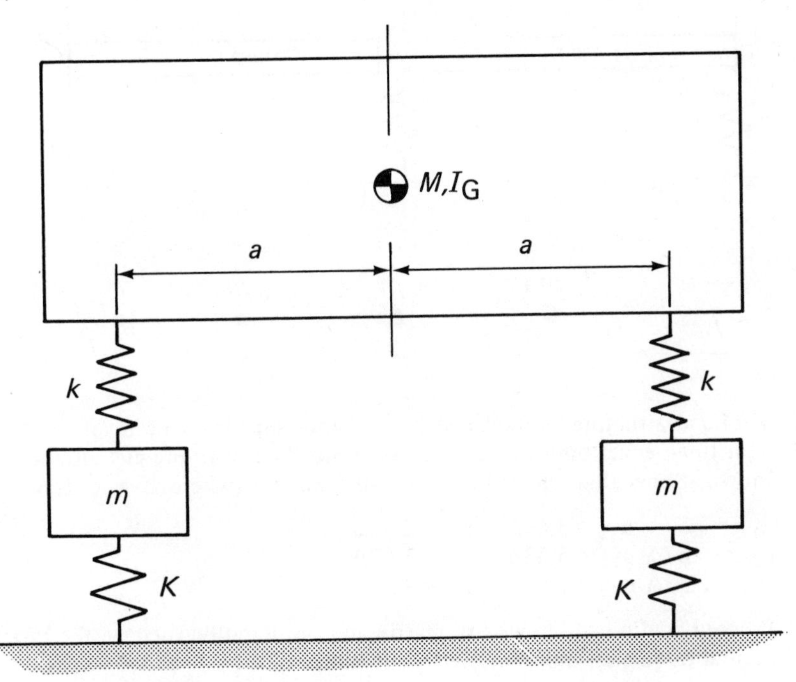

A simplified model for studying the dynamics of a motor vehicle is shown. The body has a mass M, and a moment of inertia about an axis through its mass centre of I_G. It is considered to be free to move in two directions – vertical translation and rotation in the vertical plane. Each of the unsprung wheel masses, m, are free to move in vertical translation only.
Damping effects are ignored.

(i) Derive equations of motion for this system. Define carefully the co-ordinates used.

(ii) Is it possible to determine the natural frequency of a 'wheel hop' mode without solving all the equations of motion? If not, suggest an approximation which might be made, in order to obtain an estimate of the wheel hop frequency, and calculate such an estimate given the following data:
$k = 20$ kN/m; $K = 70$ kN/m; $m = 22.5$ kg.

27. To analyse the vibration of a two coach rail unit, it is modelled as the system shown. Each coach is represented by a rigid uniform beam of length l and mass m: the coupling is a simple ball-joint. The suspension is considered to be three similar springs, each of stiffness k, positioned as shown. Damping can be neglected.
Considering motion in the plane of the figure only, obtain the equations of motion for small amplitude free vibrations, and hence obtain the natural frequencies of the system.
Explain how the mode shapes may be found.

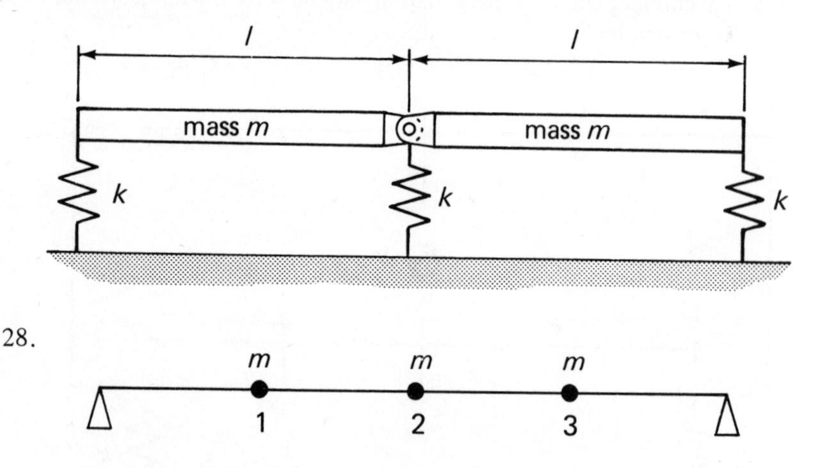

28. A bridge structure is modelled by a simply supported beam of length l, with three equal bodies each of mass m attached to it at equal distances as shown. Show that the influence coefficients are (where $\Delta = l^3/256\,EI$):

$$\alpha_{11} = 3\Delta \quad \alpha_{12} = 3.67\Delta \quad \alpha_{13} = 2.33\Delta$$
$$\alpha_{21} = 3.67\Delta \quad \alpha_{22} = 5.33\Delta \quad \alpha_{23} = 3.67\Delta$$
$$\alpha_{31} = 2.33\Delta \quad \alpha_{32} = 3.67\Delta \quad \alpha_{33} = 3\Delta$$

Proceed to find the flexibility matrix and, by iteration, deduce the lowest natural frequency and associated mode shape.

29. A solid cylinder has a mass M and radius R. Pinned to the axis of the cylinder is an arm of length l which carries a bob of mass m. Obtain the natural frequency of free vibration of the bob.

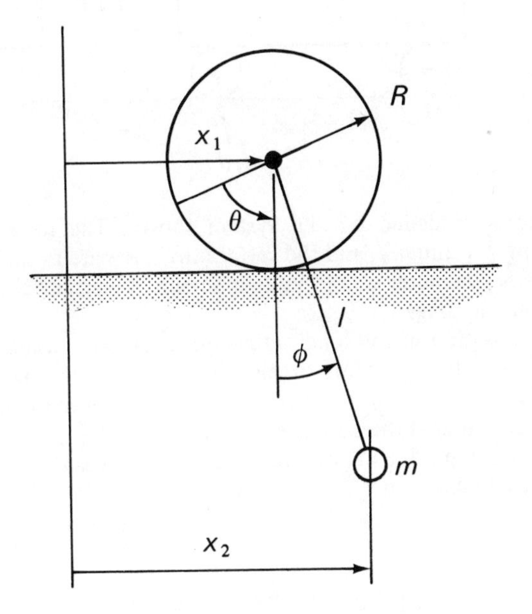

30. An aeroplane has a fuselage mass of 4000 kg. Each wing has an engine of mass 500 kg, and a fuel tank of mass 200 kg at its tip, as shown. Neglecting the mass of each wing, calculate the frequencies of flexural vibrations in a vertical plane. Take the stiffness of the wing sections to be 3 k and k as shown, where k = 100 kN/m.

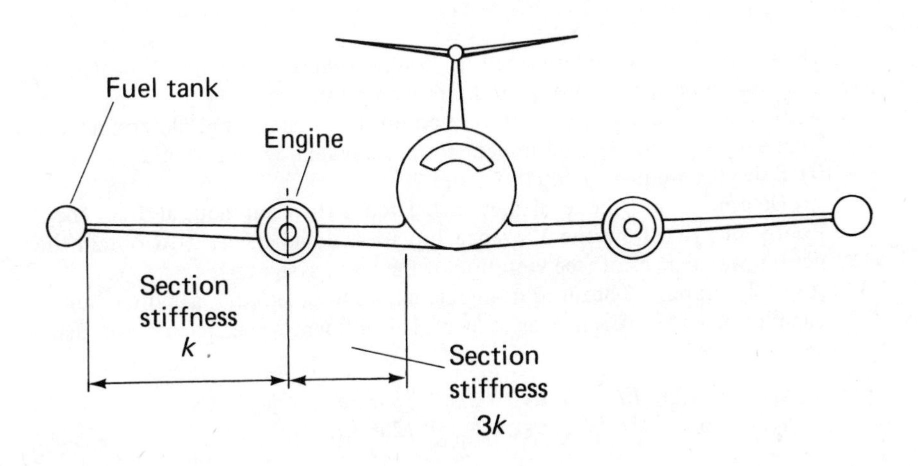

31.

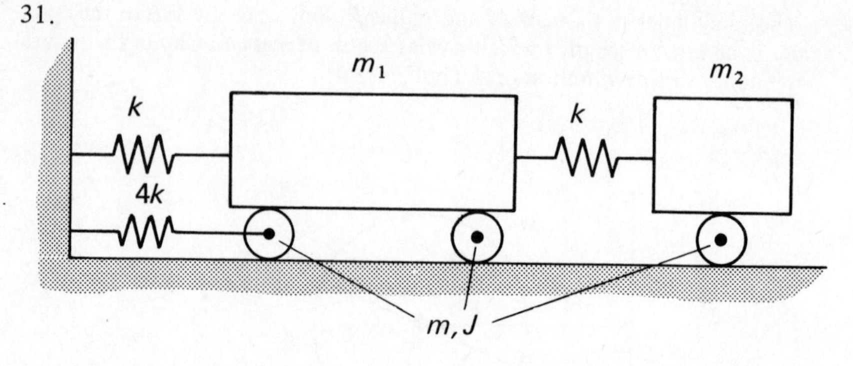

A machine is modelled by the system shown. The masses of the main elements are m_1 and m_2, and the spring stiffnesses are as shown. Each roller has a mass m, diameter d, and mass moment of inertia J about its axis, and rolls without slipping.

Considering motion in the longitudinal direction only, use Lagrange's Equation to obtain the equations of motion for small free oscillations of the system. If $m_1 = 4$ m, $m_2 = 2$ m and $J = md^2/8$, deduce the natural frequencies of the system and the corresponding mode shapes.

32. Vibrations of a particular structure can be analysed by considering the equivalent system shown.

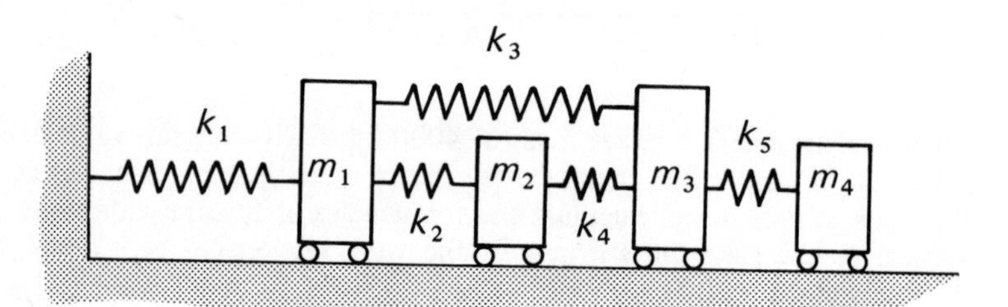

The bodies are mounted on small frictionless rollers whose mass is negligible, and motion occurs in a horizontal direction only.

Write down the equations of motion of the system and determine the frequency equation in determinant form. Indicate how you would

(i) Solve the frequency equation, and

(ii) Determine the mode shapes associated with each natural frequency.

Briefly describe how the Lagrange Equation could be used to obtain the natural frequencies of free vibration of the given system.

33. A simply supported beam of negligible mass and of length l, has three bodies each of mass m attached as shown. The influence coefficients are, using standard notation,

$$\alpha_{11} = 3l^3/256\,EI \qquad \alpha_{31} = 2.33l^3/256\,EI$$
$$\alpha_{21} = 3.67l^3/256\,EI \qquad \alpha_{22} = 5.33l^3/256\,EI.$$

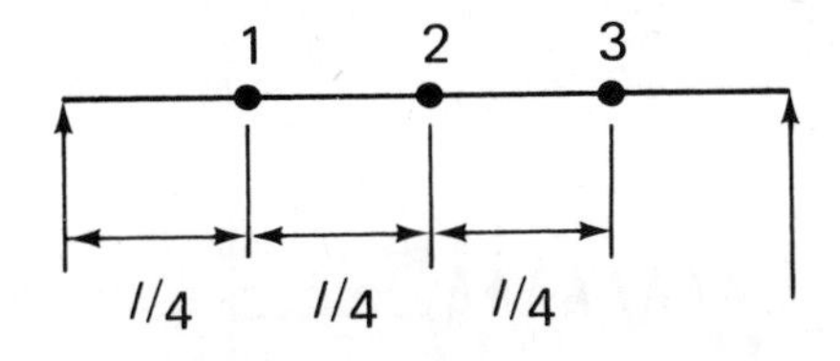

Write down the flexibility matrix, and determine by iteration the frequency of the first mode of vibration, correct to 2 significant figures, if $EI = 10$ Nm^2, $m = 2$ kg and $l = 1$ m.
Comment on the physical meaning of the eigenvector you have obtained, and use the orthogonality principle to obtain the frequencies of the higher modes.

34. A structure is modelled by three identical long beams and rigid bodies, connected by two springs as shown. The rigid bodies are each of mass M and the mass of the beams is negligible. Each beam has a transverse stiffness K at its unsupported end, and the springs have stiffnesses k and $2\,k$ as shown. Determine the frequencies and corresponding mode shapes of small amplitude oscillation of the bodies in the plane of the figure. Damping can be neglected.

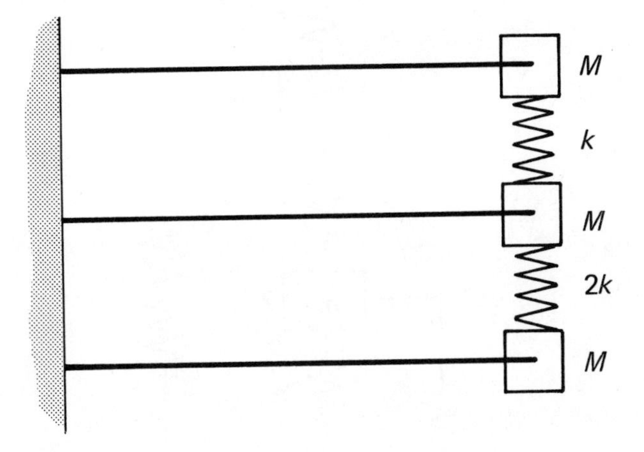

35.

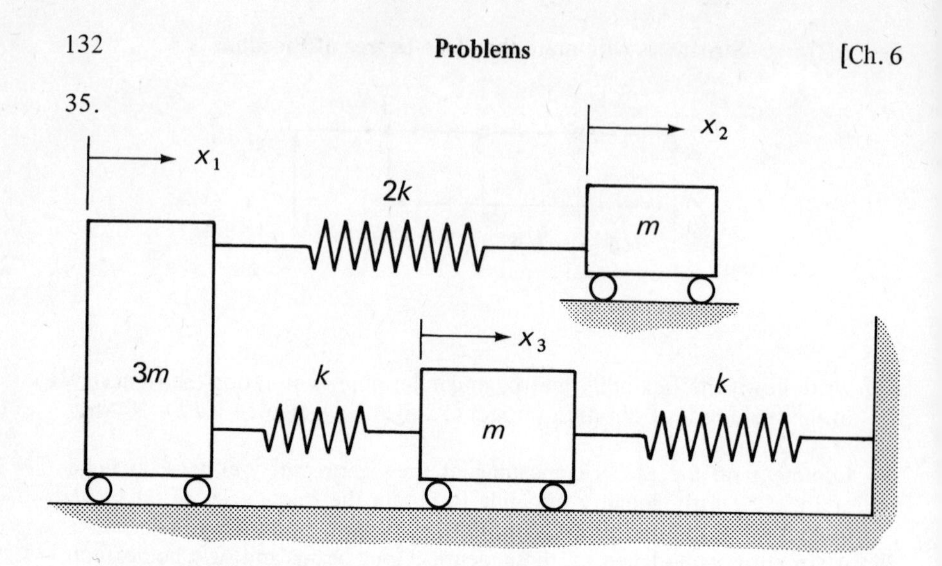

Find the dynamic matrix of the system shown.
If $k = 20$ kN/m and $m = 5$ kg, find the lowest natural frequency of the system and the associated mode shape.

36. A structure is modelled by the three degree of freedom system shown. Only translational motion in a vertical direction can occur.

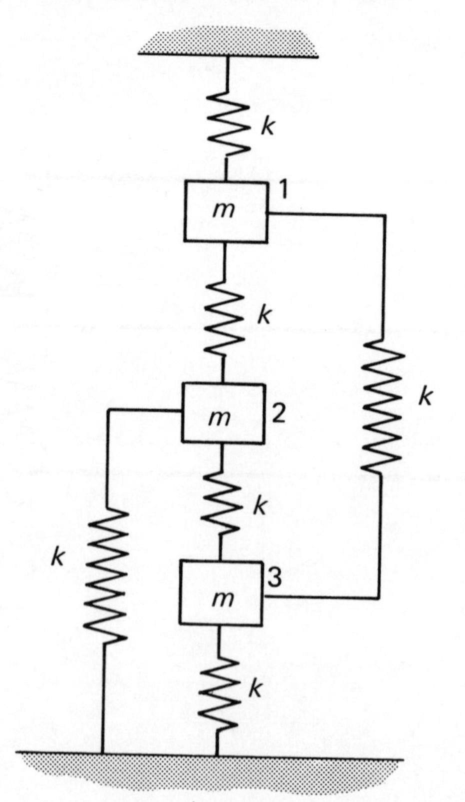

Show that the influence coefficients are

$$\alpha_{11} = \alpha_{22} = \alpha_{33} = \tfrac{1}{2}k$$

and $$\alpha_{21} = \alpha_{31} = \alpha_{32} = \tfrac{1}{4}k,$$

and proceed to find the flexibility matrix. Hence obtain the lowest natural frequency of the system and the corresponding mode shape.

37. A delicate instrument is to be mounted on an antivibration installation so as to minimise the risk of interference caused by groundborne vibration. An elevation of the installation is shown below, and the point A indicates the location of the most sensitive part of the instrument. The installation is free to move in the vertical plane, but horizontal translation is not to be considered.

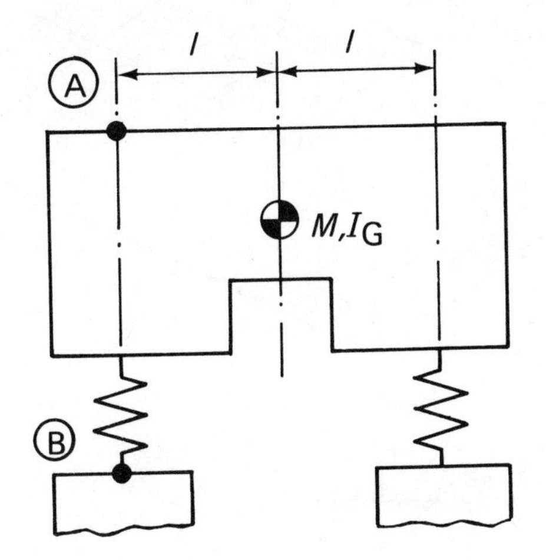

It is decided to use as a design criterion the transmissibility T_{AB}, being the sinusoidal vibration amplitude at A for a unit amplitude of vibration on the ground at B. One of the major sources of groundborne vibration is a nearby workshop where there are several machines which run at 3000 rev/min. Accordingly, it is proposed that the installation should have a transmissibility $|T_{AB}|$ of 1% at 50 Hz.

Given the following data:

$$M = 3175 \text{ kg}; \; l = 0.75 \text{ m}; \; R = 0.43 \text{ m (where } I_G = MR^2);$$

determine the maximum value of stiffness K which the mounts may possess in order to meet the requirement, and find the two natural frequencies of the installation.

Repeat the analysis using a simpler model of the system having just one degree of freedom – vertical translation of the whole installation – and establish whether this simpler approach provides an acceptable means of designing such a vibration isolation system.

For the purpose of these basic isolation design calculations, damping may be ignored.

38. In a vibration isolation system, a group of machines are firmly mounted together onto a rigid concrete raft which is then isolated from the foundation by 4 antivibration pads. For purposes of analysis, the system may be modelled as a symmetrical body of mass 1150 kg and moments of inertia about rolling and pitching axes through the mass centre of 175 kgm^2 and 250 kgm^2 respectively, supported at each corner by a spring of stiffness 7.5×10^5 N/m. The model is shown below.

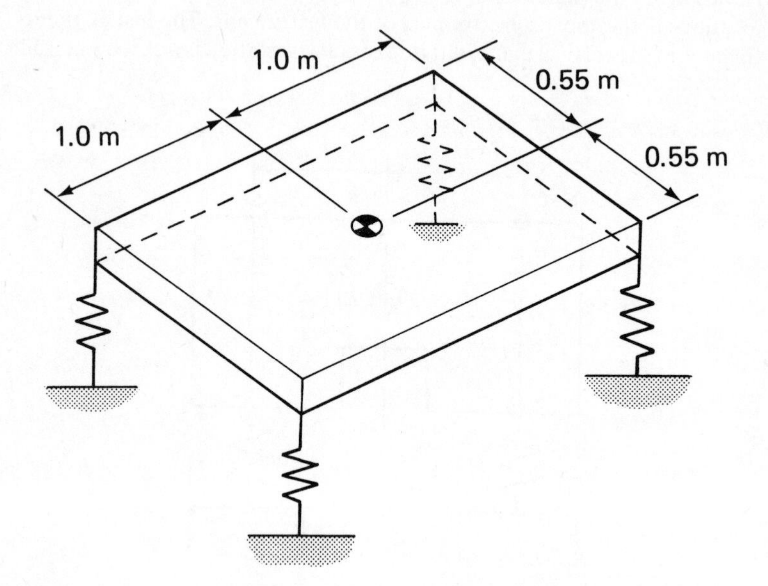

The major disturbing force is generated by a machine at one corner of the raft and may be represented by a harmonically varying vertical force with a frequency of 50 Hz, acting directly through the axis of one of the mounts.

(i) Considering vertical vibration only, show that the force transmitted to the foundation by each mount will be different, and calculate the magnitude of the largest, expressed as a percentage of the excitation force.

(ii) Identify the mode of vibration which is responsible for the largest component of this transmitted force and suggest ways of improving the isolation performance using the same mounts but *without* modifying the raft.

(iii) Show that a considerable improvement in isolation would be obtained by moving the disturbing machine to the centre of the raft, and calculate the transmitted force for this case, again expressed as a percentage of the exciting force.

39. Find the driving point impedance of the system shown. The bodies move on frictionless rollers in a horizontal direction only.

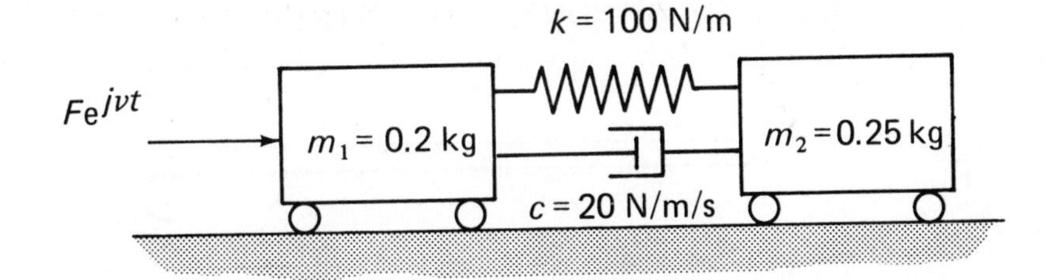

Hence show that the amplitude of body 1 is

$$\frac{\sqrt{(72\,000 + 2620\nu^2 + 0.2\nu^4)^2 + (20\nu^3)^2}}{\nu^2\,(0.04\nu^4 + 1224\nu^2 + 32\,400)}\,F.$$

40. Find the driving point mobility of the system shown, only motion in the vertical direction occurs and damping is negligible.

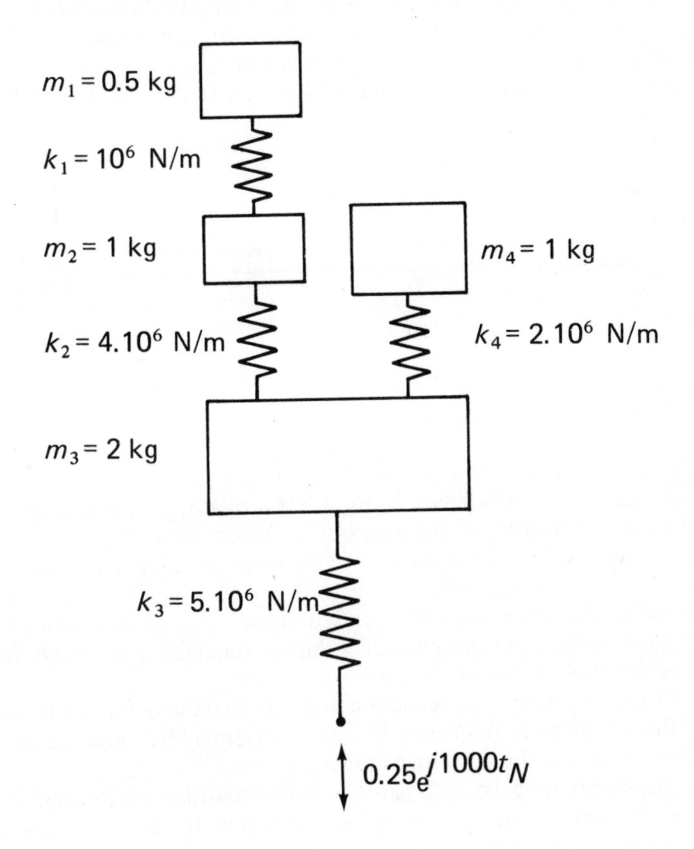

Hence obtain the frequency equation: check your result by using a different method of analysis.

6.3 THE VIBRATION OF CONTINUOUS STRUCTURES

41.

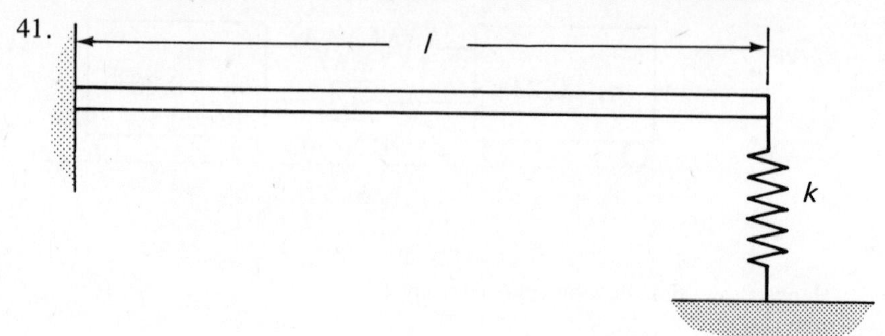

A uniform beam of length l is built-in at one end, and rests on a spring of stiffness k at the other, as shown.
Determine the frequency equation for small amplitude transverse vibration, and show how the first natural frequency changes as k increases from zero, a free end, to infinity, a simply supported end.
Comment on the effect of the value of k on the frequency of the 10th mode.

42.

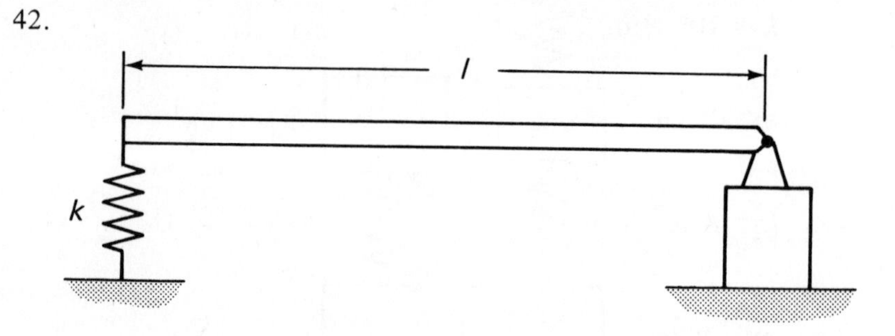

A structure is modelled as a uniform beam of length l, hinged at one end, and resting on a spring of stiffness k at the other, as shown.
Determine the first three natural frequencies of the beam, and sketch the corresponding mode shapes.

43. Part of a structure is modelled as a uniform cross-section beam having a pinned attachment at one end and a sliding constraint at the other (where it is free to translate, but not to rotate) as shown,
    (i) Derive the frequency equation for this beam and find expressions for the nth natural frequency and the corresponding mode shape. Sketch the shapes of the first three modes.
    (ii) The beam is to be stiffened by adding a spring of stiffness k to the sliding end. Derive the frequency equation for this case and use the result to deduce the frequency equation for a pinned clamped beam.
    (iii) Estimate how much the fundamental frequency of the original beam is raised by adding a very stiff spring to its sliding end.

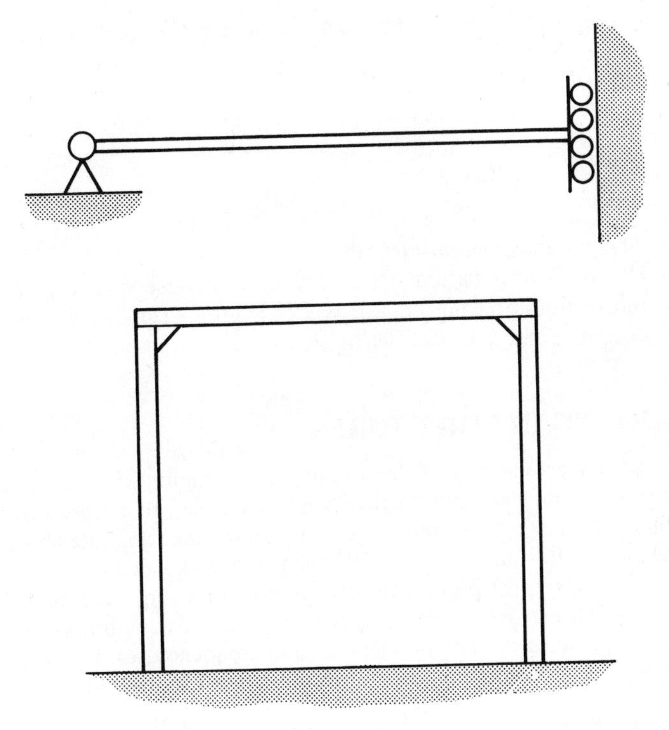

44. A portal frame consists of three uniform beams, each of length l, mass m, and flexural rigidity EI, attached as shown. There is no relative rotation between beams at their joints.
Show that the fundamental frequency of free vibration, in the plane of the frame, is $0.5\sqrt{EI/ml^3}$ Hz.

45. A uniform cantilever of length l and flexural rigidity EI, is subjected to a transverse harmonic exciting force $F \sin \nu t$ at the free end. Show that the displacement at the free end is

$$\left[\frac{\sin \lambda l . \cosh \lambda l - \cos \lambda l . \sinh \lambda l}{EI\lambda^3 \; (1 + \cos \lambda l . \cosh \lambda l)}\right] F \sin \nu t,$$

where $\lambda = (\rho A \nu^2 / EI)^{1/4}$.

46. A thin rectangular plate has its long sides simply supported, and both its short sides unsupported. Find the first three natural frequencies of flexural vibration, and sketch the corresponding mode shapes.

47. Derive the frequency equation for flexural vibration of a uniform beam which is pinned at one end and free at the other.
Show that the fundamental mode of vibration has a natural frequency of zero, and explain the physical significance of this mode.
Obtain an approximate value for the natural frequency of the first bending mode of vibration, and compare this with the corresponding value for a beam which is rigidly clamped at one end and free at the other.

48. Part of the cooling system in a generating station consists of a steel pipe 80 mm in outer diameter, 5 mm thick and 4 m long. The pipe may be assumed to be built in at each end; show that the static deflection y, at a distance x from one end of the pipe of length l, is given by

$$y = \frac{mg}{24\,EI}\,x^2\,(l-x)^2\,,$$

where m is the mass per unit length.
Calculate the lowest natural frequency of transverse vibration of the pipe when full of liquid having a density of 930 kg/m^3. Take the density of steel as 7750 kg/m^3, and E as 200 GN/m^2.

6.4 DAMPING IN STRUCTURES

49. The 'half-power' method of determining the damping in a particular mode of vibration from a receptance plot, can be extended to a more general form in which the two points used – one below resonance and one above – need not be at an amplitude exactly 0.707 times the peak value.
(i) A typical Nyquist plot of the receptance for a single degree of freedom system with structural damping is shown, with two points corresponding to frequencies ν_1 and ν_2. The natural frequency, ω, is also indicated. Prove that the damping loss factor, η, is given exactly by:

$$\eta = [(\nu_2{}^2 - \nu_1{}^2)/\omega^2]\;[(\tan\tfrac{1}{2}\phi_1 + \tan\tfrac{1}{2}\phi_2)]^{-1}\,,$$

where ϕ_1 and ϕ_2 are the angles subtended by points 1 and 2 with the resonance point and the centre of the circle. Show how this expression relates to the half-power points formula.

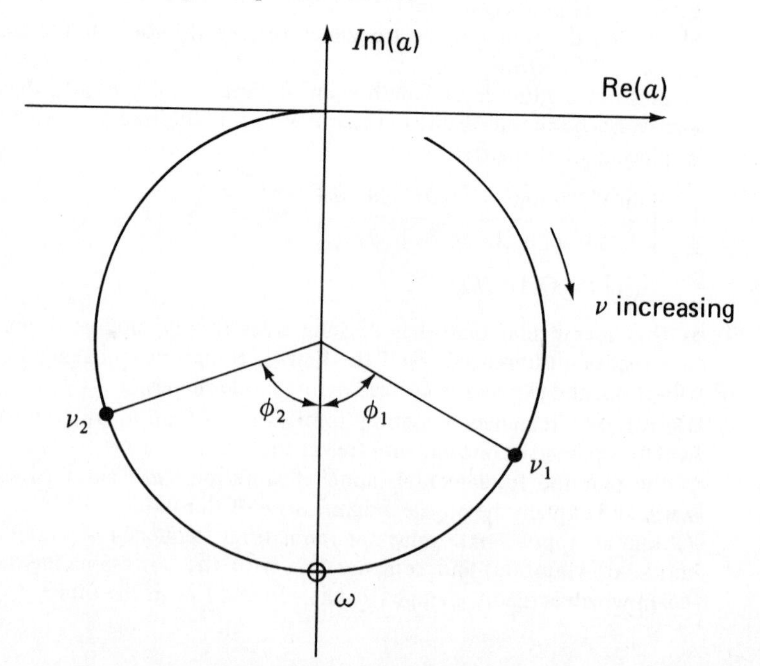

(ii) Some receptance data from measurements on a practical structure are listed in the table. By application of the formula above to the data given, obtain a best estimate for the damping of the mode under investigation.

Frequency (Hz)	Receptance	
	Modulus ($\times 10^{-7}$ m/N)	Phase (degrees)
5.86	12.3	24
5.87	12.6	29
5.88	12.5	36.5
5.89	12.0	41
5.90	11.3	57
5.91	10.1	66
5.92	8.8	74
5.93	7.0	78
5.94	5.6	78

50. A large symmetric machine tool structure is supported by four suspension units, one at each corner, intended to provide isolation against vibration. Each unit consists of a primary spring (which can be considered massless and undamped) of stiffness 250 kN/m, in parallel with a viscous dashpot of rate 20 kN s/m. The 'bouncing' natural frequency of the installation is 2.8 Hz while the two rocking modes are 1.9 Hz and 2.2 Hz.
It is found that excessive high frequency vibration forces are transmitted from the machine to the floor, particularly above 20 Hz. Some modifications are required to improve the isolation performance, but a constraint is imposed by the pipes and other service connections to the machine which cannot withstand significantly larger displacements than are currently encountered.
It is suggested that a rubber bush be inserted *either* at one end of the dashpot (e.g. between the dashpot and the machine structure), *or* between the entire suspension unit and the machine. The same bush would be used in either configuration and it may be modelled as an undamped spring with a stiffness of 700 kN/m.
Show analytically which of the two proposed modifications provides the greatest improvement in high frequency isolation, and calculate the increased attenuation (in dB) for both cases at 25 Hz and at 50 Hz. Consider motion in the vertical direction only.
Comment on the suitability of the two proposed modifications, and indicate what additional calculations should be made to define completely the dynamic behaviour of the modified installation.
51. A machine having a mass of 1250 kg is isolated from floor vibration by a resilient mount whose stiffness is 0.2×10^6 N/m, and which has negligible damping. The machine generates a strong excitation which can be considered as an externally applied harmonic force at its running speed of 480 revolutions per minute, and the vibration isolation required is specified as a force transmissibility of -35 dB at this frequency.

(i) Show that the single-stage system described above will not provide the necessary attenuation.
It is decided to improve the effectiveness of the installation by introducing a second mass-spring stage between the resilient mount and the floor. The maximum deadweight which can be supported by the floor is 2500 kg, and so the second stage mass is taken as 1250 kg.
(ii) Calculate the stiffness of the second-stage spring in order to attain the required force transmissibility.
(iii) Determine the frequency at which this two-stage system has the same transmissibility as the simpler single-stage one, and sketch the transmissibility curve for each case, indicating the frequency above which the isolation system gives a definite attenuation.

52. (i) The traditional 'Half-power points' formula for estimating damping: loss factor $= \Delta f / f_0$, (where Δf is the frequency bandwidth at the half power points and f_0 is the frequency of maximum response) is an approximation which becomes unreliable when applied to modes with relatively high damping. Sketch a graph indicating the error incurred in using this formula instead of the exact one, as a function of damping loss factor in the range 0.1 to 1.0.
(ii) The measured receptance data given in the table were taken in the frequency region near a mode of vibration of interest on a scale model of a chemical reactor. Obtain estimates for the damping in this mode of vibration using (*a*) a modulus-frequency plot and (*b*) a polar (or Nyquist plot of the receptance data. Present your answers in terms of Q factors. State which of the two estimates obtained you consider to be the more reliable, and justify your choice.

Frequency (Hz)	Receptance	
	Modulus ($\times 10^{-6}$ m/N)	Phase (degrees)
380.0	41.6	31
390.0	49.9	25
400.0	66.5	25
410.0	86.1	41
420.0	70.7	67
430.0	64.0	65
440.0	67.0	60

53. The results are given below of an incomplete resonance test on a structure. The response at different frequencies was measured at the point of application of a sinusoidal driving force and is given as the receptance, being the ratio of the amplitude of vibration to the maximum value of the force. The phase angle between the amplitude and force was also measured.
Estimate the effective mass, dynamic stiffness and the loss factor, assuming material type damping.

Frequency Hz	Receptance ($\times 10^{-6}$ m/N)	Phase angle (degrees)
55	5	0
70	10	3
82	18	24
88	25	40
94	30	55
100	32	85
109	10	135
115	9	180
130	7	180

54. A test is conducted in order to measure the dynamic properties of an anti-vibration mount. A mass of 900 kg is supported on the mount to form a single-degree-of-freedom system, and measurements are made of the receptance of this system in the region of its major resonance.
In addition to the hysteretic damping provided by the mount (and which is to be measured), some additional damping is introduced by friction in the apparatus, and so tests are made at two different amplitudes of vibration (x_o) in order to determine the magnitude of each of the two sources of damping.
It may be assumed that the loss factor of the mount is a constant, valid for all vibration amplitudes, but the dynamic stiffness is not a constant and so the two tests have slightly different natural frequencies.
Details of some receptance measurements are given below. Assuming that the additional damping has the characteristic of coulomb friction damping, estimate the hysteretic damping loss factor of the mount.

Receptance measurements

Frequency (Hz)	Test (a) $x_o = 0.1$ mm		Test (b) $x_o = 0.02$ mm	
	Re(α) ($\times 10^{-7}$ m/N)	Im(α) ($\times 10^{-7}$ m/N)	Re(α) ($\times 10^{-7}$ m/N)	Im(α) ($\times 10^{-7}$ m/N)
13.25	7.6	–6.9	5.1	–4.9
13.50	7.6	–8.6	5.3	–5.6
13.75	6.6	–10.7	5.3	–6.4
14.00	4.4	–12.4	5.2	–7.4
14.25	1.6	–12.8	4.6	–8.5
14.50	–1.0	–11.9	3.7	–9.6
14.75	–2.5	–10.1	2.2	–10.3
15.00	–3.1	–8.4	0.58	–10.2
15.25	–3.2	–7.0	–0.96	–9.6
15.50	–3.0	–6.0	–1.9	–8.5

55. A resonance test on a flexible structure at a constant energy level has revealed a prominent mode at 120 Hz with a half power frequency bandwidth of 4.8 Hz and a peak acceleration of 480 m/s^2. The effective mass of the structure for this mode has been estimated at 20 kg.
    It is proposed to introduce coulomb type friction at the point of measurement of the response so as to reduce the motion by 1/5 for the same energy input as before. Estimate the friction force required, assuming this to be independent of amplitude and frequency of vibration.
56. (i) Derive a relationship between the logarithmic decrement of a system with velocity type damping performing free vibrations, and the loss factor for structural damping. Define clearly any assumptions made.
    (ii) A concrete floor slab is to be supported on four columns, spaced in a square grid of sides 7 m. The detail of one column is shown. The slab is to be isolated from vibrations being transmitted up the column by rubber pads, installed as shown. The slab thickness is 150 mm and the density of concrete is 2250 kg/m^3. The first resonance frequency of the slab is estimated to be 20 Hz and the logarithmic decrement for concrete is about 0.2. Measurements of the vibration in a column have shown a strong peak at 20 Hz. The rubber pads have a loss factor of 0.01.
    (a) Estimate the resonance frequency the pad system should have to provide a reduction of 4/5 in the vibration being transmitted at 20 Hz to the centre of the floor slab. Comment upon the result.
    (b) Estimate the additional attenuation in dB's the isolation will provide at a frequency of 160 Hz.

    Some of the information given may be superfluous.

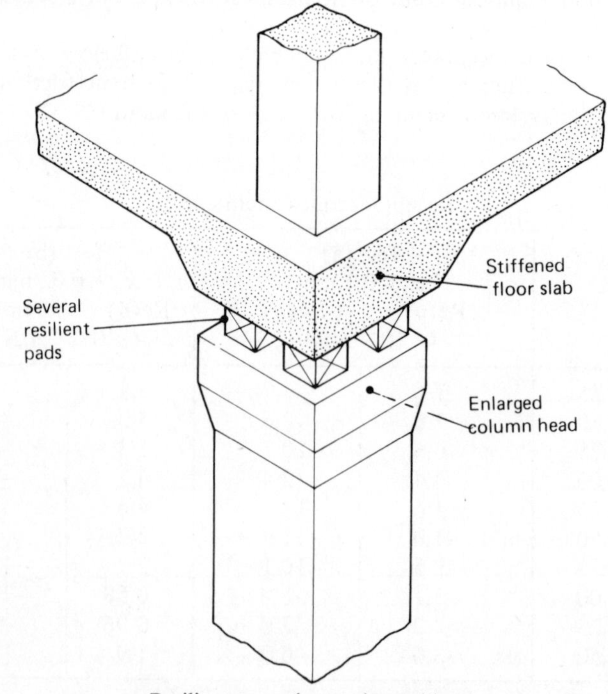

Resilient mounting at elevated level

57. (i) Often it is required to introduce into a structure some additional form of damping in order to keep resonance vibrations down to an acceptable level. One method is to use a damped dynamic absorber where a (relatively) small mass is suspended from the primary mass (of the vibrating structure) by a spring and a dashpot. If the absorber spring-mass system is tuned to the natural frequency of the effective mass-spring model of the structure, then the absorber dashpot may introduce some damping to the structural resonance.
Without performing analysis, but using physical reasoning only, sketch a family of curves for the point receptance on the main mass for a range of different magnitudes for the absorber dashpot between 0 and ∞, and hence show that there will be an optimum value for that dashpot rate.
(ii) In one application of this type of damper-absorber, it is required to increase the damping in a new suspension bridge.
In moderate to high winds the airflow over the bridge generates an effectively steady-state excitation force at the bridge's fundamental natural frequency. The airflow also provides some damping. The amplitude of steady vibration under this excitation is found to be 20 mm and this is twice the maximum amplitude considered to be "safe". Accordingly, it is proposed to introduce extra damping to reduce the resonance amplitude to 10 mm.
Tests on the bridge show that it possesses some structural damping and this is estimated from measurement of free decay curves. The amplitude of vibration is found to halve after 40 cycles. A more significant source of damping is the airflow over the bridge and this is most readily described in terms of energy dissipation and this is estimated to be (βx_0^3) Nm per cycle where $\beta = 2.5 \times 10^7$ N/m^2 and x_0 = vibration amplitude. The effective mass and stiffness of the bridge (for its fundamental mode) are 500,000 kg and 5×10^6 N/m respectively. Determine the equivalent viscous dashpot rate which must be added in order to reduce the resonance vibration amplitude to 10 mm. Assume the excitation force remains the same.
(iii) If, due to miscalculation, the actual dashpot used has a rate of only 70% of that specified, what then will be the vibration amplitude?

58. A partition is made from several layers of metal and a plastic material. Experiments with the metal layers alone have shown that the energy dissipated per cycle of vibration at the lowest natural frequency is $3 \times 10^4\ x_0^2$ joule/cycle, where x_0 is the amplitude in metres at the centre of the partition. The stiffness of the plastic layers themselves when measured at the centre is 4×10^5 N/m with a loss factor of 0.3. The acoustic energy loss from one face alone of the partition when vibrating at the lowest natural frequency of 70 Hz is $1.5\dot{x}^2$ joule/cycle, where $\dot{x}$ is the maximum velocity in m/s at the centre.
Calculate the amplitude of vibration at the centre of the partition when one face receives an acoustic energy input of 50 watts at 70 Hz. Explain carefully any assumptions which have to be made.

59. A sketch is given below of the essential parts of the front suspension of a motor car, showing the unsprung mass consisting of the tyre, the wheel and the stub axle, connected at point A by a rubber bush to a hydraulic shock

absorber and the main coil spring. The other end of the shock absorber is connected at point B by another rubber bush to a subframe of the car body. A set of wishbone link arms with rubber bushes at each end serve to stabilise the unit.

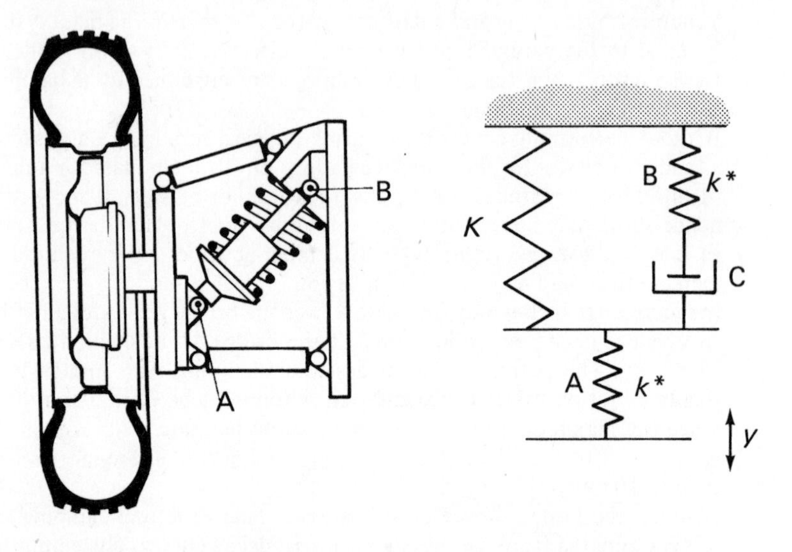

(i) Devise a representative model for this suspension system comprising lumped masses, springs and dampers. Indicate how the equations of motion can be obtained but do *not* solve these. Define the symbols introduced carefully.

(ii) A massless model of the main spring, the shock absorber and the bushes at points A and B is shown, assuming that the car body represents an infinite impedance. The rubber bushes A and B are identical and have a complex stiffness.

$$k^* = k(1 + j\eta),$$

where the elastic stiffness $k = 600$ kN/m and the loss factor $\eta = 0.25$.

The main spring stiffness $K = 25$ kN/m and the shock absorber behaves as a viscous damper with a coefficient $c = 3$ kN s/m. Estimate the percentage contribution by the two bushes to the total energy being dissipated per cycle, for an input motion $y = y_0 \sin \nu t$ with $y_0 = 25$ mm and $\nu = 30$ rad/s, and assuming that the maximum possible displacement of 5 mm across each bush is being taken up.

60. A machine produces a vertical harmonic force and is to be isolated from the foundations by a suspension system consisting of metal springs in series with blocks of a viscoelastic material.

(i) Show analytically whether it will be better to place the blocks of visco-elastic material above or below the springs from the point of view of:

(*a*) the force transmitted to the foundation

(*b*) damping out high frequency resonances in the metal spring for the type of installation where the attachment points to the machine are slender metal brackets.

(ii) Compare the system described above with one in which the blocks of viscoelastic material are placed in parallel with the springs.
Define carefully all symbols introduced.

61. A fabricated steel mast is observed to oscillate violently under certain wind conditions. In order to increase the damping some relative motion is to be allowed in a number of the bolted joints by inserting spring washers under the nuts and by opening the holes to give a definite clearance. Rubber blocks are to be provided to keep the joint central.
In 15 joints metal to metal sliding friction is to be introduced with a coefficient of friction of 0.2 for a clamping force of 2×10^4 N. The clearance in each bolt hole is 2.5 mm on diameter. To keep the joint nominally at its central position two rubber blocks are fitted as shown below. The blocks are pressed in position to provide a centering force in excess of the static friction force. Each block is square in cross section, 60 mm by 60 mm and nominally 18 mm thick. The rubber has a loss factor of 0.12.
The maximum energy input per cycle of oscillation of the mast by the wind is estimated as 1500 joules.
(i) Calculate the modulus of elasticity for the rubber material so that the full clearance in the bolt holes of all the joints is just taken up during an oscillation under the maximum wind excitation, neglecting any structural damping in the mast itself.
(ii) Estimate the Q factor for the mast with the damping in the joints, given that the stored energy in the structure is 2500 joules for a deflection which takes up the total clearance in each joint.

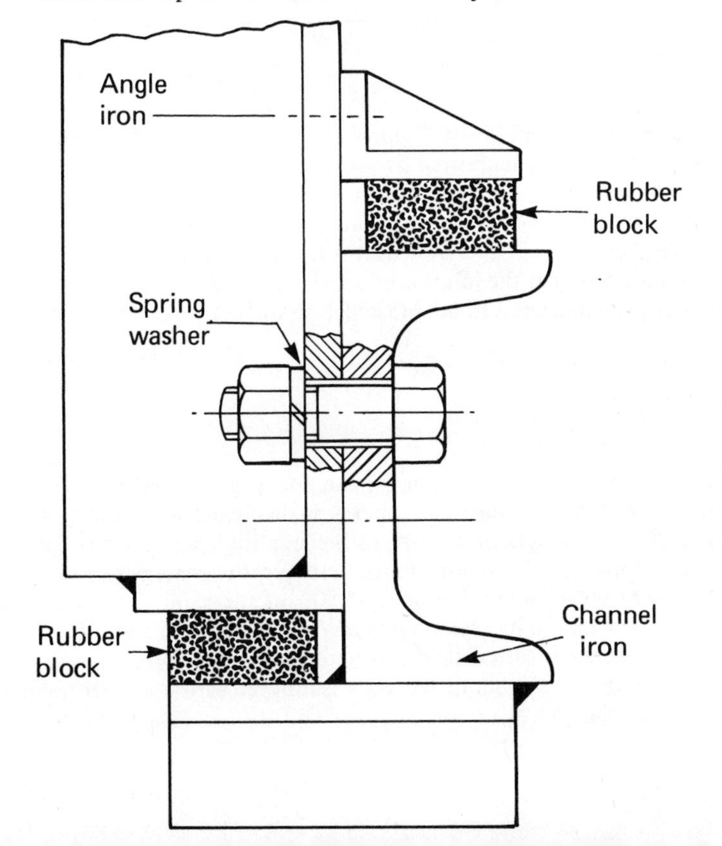

62. The receptance at a point in a structure is measured over a frequency range, and it is found that a resonance occurs in the excitation range. It is therefore decided to add an undamped vibration absorber to the structure.
Sketch a typical receptance-frequency plot for the structure, and by adding the receptance plot of an undamped vibration absorber, predict the new natural frequencies. Show the effect of changes in the absorber mass and stiffness, on the natural frequencies, by drawing new receptance-frequency curves for the absorber.
63. Consider the simple joint shown, in which metal to metal contact occurs.

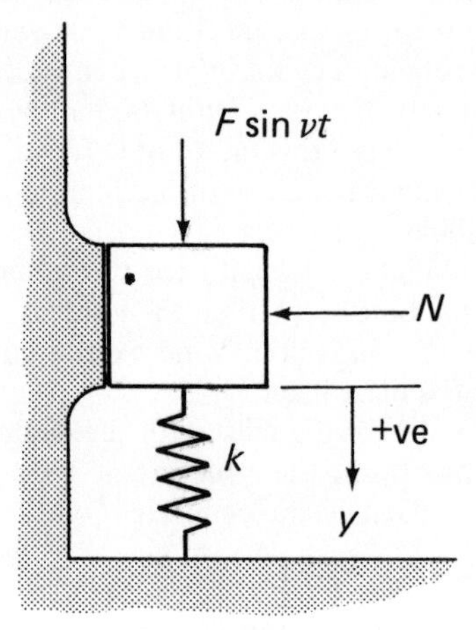

An harmonic exciting force $F \sin \nu t$ is applied to one joint member which has a mass m, and is supported by an element of stiffness k. The other member is rigidly fixed, so that it is infinitely stiff in the direction of this exciting force. A constant force N is applied normal to the joint interfaces by a clamping arrangement not shown. It is to be assumed that the coefficient of friction μ existing at the joint interfaces is constant.
Assuming the motion y to be sinusoidal, show that

$$y = \frac{F \sin \nu t - \mu N}{k\left(1 - \left(\frac{\nu}{\omega}\right)^2\right)},$$

and hence obtain an expression for the energy dissipated per cycle by slipping. Show that the maximum energy is dissipated when $\mu N/F = 0.5$, and that y then has an amplitude 50% of the amplitude when $N = 0$. Further, by drawing force-slip hysteresis loops and plotting energy dissipation as a function of $\mu N/F$, show that at least 50% of the maximum energy dissipation can be achieved by maintaining $\mu N/F$ between 0.15 and 0.85.
Comment on the practical significance of this.
64. Part of a structure is modelled by a cantilever with a friction joint at the free end, as shown.

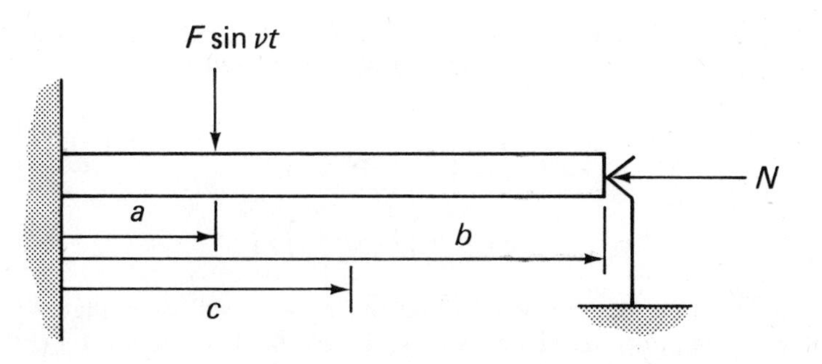

The cantilever has an harmonic exciting force $F \sin \nu t$ applied at a distance a from the root. The tangential friction force generated in the joint by the clamping force N can be represented by a series of linear periodic functions, $F_d(t)$.
Show that $y_c(t) = \alpha_{ca} F \sin \nu t + \alpha_{cb} F_d(t)$, where c is an arbitrary position along the cantilever and α is a receptance.
By assuming that the friction force is harmonic and always opposes the exciting force, find the energy dissipated per cycle, and hence show that $F_d = 2\mu N$. Is this assumption reasonable for all modes of vibration?
Thus, this linearisation of the damping replaces the actual friction force during slipping, μN, by a sinusoidally varying force of amplitude $2\ \mu N$. Compare this representation with a Fourier series for the friction damping force.

65. A beam on elastic supports with friction damped joints is modelled by the system shown.

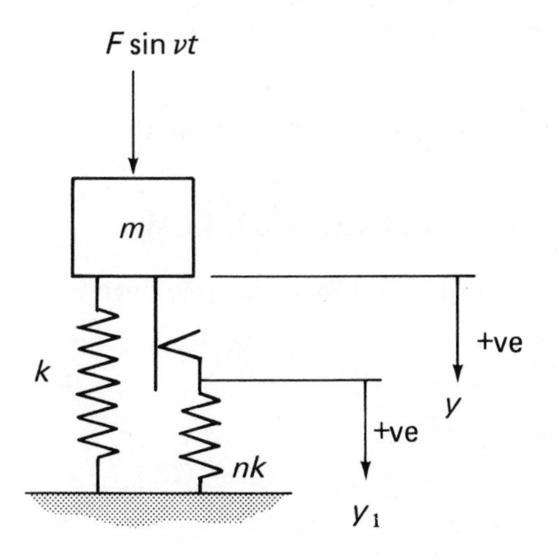

By considering equivalent viscous damping for the friction damper, show that

$$|\delta|^2 = Y^2 - (4F_d/\pi nk)^2,$$

where $\delta = y - y_1$, Y is the amplitude of the body motion, and F_d is the tangential friction force in the damper. Hence deduce that

$$Y = \left[\frac{\left(\frac{F}{k}\right)^2 + \left(\frac{4F_d}{\pi nk}\right)^2 \left\{ \left[1 - \left(\frac{\nu}{\omega}\right)^2\right]^2 - \left[1 + n - \left(\frac{\nu}{\omega}\right)^2\right]^2 \right\}}{\left[1 - \left(\frac{\nu}{\omega}\right)^2\right]^2} \right]^{1/2}$$

Consider the response when $F_d = 0$ and $F_d = \infty$, and show that the amplitude of the body for all values of F_d is $2F/nK$ when $\nu/\omega = \sqrt{1 + (n/2)}$, and assess the significance of this.

Hint: Write equations of motion for system with equivalent viscous damping $c = 4F_d/\pi\nu|\delta|$, and put $y_1 = Y_1 e^{j\nu t}$ etc. From equations of motion,

$$Y = \frac{F}{k} \left[\frac{1 + (c\nu/nk)^2}{\left[1 - \left(\frac{\nu}{\omega}\right)^2\right]^2 + \left(\frac{c\nu}{nk}\right)^2 \left[1 + n - \left(\frac{\nu}{\omega}\right)^2\right]^2} \right]^{1/2}.$$

Substituting for c and $|\delta|$ gives required expression for Y. Note that as $F_d \to \infty$, $|\delta| \to 0$.

66. Steel chimneys may be prone to appreciable wind-excited swaying oscillations due to vortex shedding. One criterion for the desirable level of damping in a chimney to prevent wind-induced oscillation is $2\,m\Lambda/\rho D^2 > 17$, where m is the mass/unit length and D the diameter of the chimney, ρ the air density and Λ is the logarithmic decrement.
For a particular chimney $m/\rho D^2 \triangleq 200$, and $\Lambda = 0.030$, so $2m\Lambda/\rho D^2 = 12$; that is the inherent damping in the chimney is insufficient to prevent wind-induced oscillations, and an increase of at least 40% is required.
What methods are available for
(i) increasing the damping in an existing chimney, and
(ii) increasing the damping in a chimney which is still at the design stage.

6.5 ANSWERS TO SELECTED PROBLEMS

6.5.1 The vibration of structures with one degree of freedom

1. 5.8 Hz.
2. $\frac{1}{2\pi}\sqrt{\frac{12kl - 3mg}{2ml}}$ Hz.
3. 3.6 Hz.
4. 25 Hz.
5. 1.64 Hz.
6. 1.45 Hz.
7. 8.5 Hz.
8. 9.8 Hz.
10. 5.5 s; 65 m.

12. 0.997 s.
13. 10 600 Ns/m; 0.035; 0.0056.
14. 1.7 mm.
15. 59 200 kg.
16. 5.3 kN/m.
17. 0.056 m; 2.25°.
18. $f(t) = -\dfrac{8}{\pi^2}\left[\cos t + \dfrac{1}{9}\cos 3t + \dfrac{1}{25}\cos 5t + \dots\ .\right]$

6.5.2 The vibration of structures with more than one degree of freedom

20. $m^2\omega^4 - 3mk\omega^2 + k^2 = 0;\ \sqrt{\dfrac{k}{m}}\ \sqrt{\dfrac{3 \pm \sqrt{5}}{2}}.$
21. 1.1 Hz, 4.35 Hz; –3.15, 0.079; 78 km/h, 19.8 km/h.
22. $\dfrac{1}{2\pi}\sqrt{\dfrac{k_1}{2m_1}}$ Hz, +0.5; $\dfrac{1}{2\pi}\sqrt{\dfrac{2k_1}{m_1}}$ Hz, –1.0.
23. $\dfrac{1}{2\pi}\sqrt{\dfrac{2k}{m}}$ Hz; $\dfrac{1}{2\pi}\sqrt{\dfrac{8k}{m}}$ Hz.
26. 10.1 Hz.
27. $\dfrac{1}{2\pi}\sqrt{\dfrac{3k}{m}}$ Hz; $\dfrac{1}{2\pi}\sqrt{\dfrac{3k \pm \sqrt{3k}}{m}}$ Hz.
29. $\dfrac{1}{2\pi}\sqrt{\left(1 + \dfrac{2m}{3M}\right)\dfrac{g}{l}}$ Hz.
30. 3.09 Hz; 5.22 Hz.
31. $\dfrac{0.459}{2\pi}\sqrt{\dfrac{k}{m}}$ Hz; $\dfrac{0.918}{2\pi}\sqrt{\dfrac{k}{m}}$ Hz; –1.0.
35. $\begin{bmatrix} k/m & -2k/3m & -k/3m \\ -2k/m & 2k/3m & 0 \\ -k/m & 0 & 2k/3m \end{bmatrix}$
37. 775 kN/m; 3.52 Hz, 6.13 Hz. Unacceptable, $k = 1570$ kN/m.
38. 5.5%; 0.68%.

6.5.3 The vibration of continuous structures

48. 25.6 Hz.
50. At one end, 10 dB and 16 dB; between 13 dB and 19 dB.
52. 14, 19.
53. 0.12.

ACKNOWLEDGEMENT

Some of the problems first appeared in the University of London B.Sc. (Engineering) Part 3 Degree Examination in Vibrations, set for students at Imperial College of Science and Technology.

Index